The Risk Analysis Controversy
An Institutional Perspective

Proceedings of a Summer Study on Decision Processes and Institutional Aspects of Risk held at IIASA, Laxenburg, Austria, 22-26 June, 1981

Editors:
Howard C. Kunreuther and Eryl V. Ley

Springer-Verlag Berlin Heidelberg New York 1982

Howard C. KUNREUTHER
Department of Decision Sciences
Wharton School
University of Pennsylvania
Philadelphia, PA 19104

Eryl V. LEY
International Institute for Applied
Systems Analysis
A-2361 Laxenburg
Austria

ISBN-13:978-3-642-81942-1 e-ISBN-13:978-3-642-81940-7
DOI: 10.1007/978-3-642-81940-7

Library of Congress Cataloging in Publication Data
Main entry under title:
The Risk analysis controversy.
Includes index.
1. Technology assessment-Congresses. 2. Risk-Congresses. I. Kunreuther, Howard. II. Ley, Eryl V.
III. Internation Institute for Applied Systems Analysis.
T174.5.R554 1982 363.1'063 82-16982

Softcover reprint of the hardcover 1st edition 1982

2061/3020 – 543210

Preface

The first summer study at IIASA brought together a cross-section of individuals from different disciplines and nationalities. All the participants have had an interest in the role of risk analysis given the institutional arrangements which guide decision making for new technologies. This book contains edited versions of the papers presented at the meeting as well as a transcript of the discussions which took place. It provides the ingredients for a broader framework fcr studying the problems associated with technology and society where risk is representative of a much wider set of concerns than simply the probability and consequences of a hazardous accident.

The Bundesministerium fuer Forschung und Technologie has an interest in promoting risk and safety research because of these new developments in society over the past ten years. In particular, there has been a diminished confidence in experts' statements on risk and a realization that many of the events which are being examined are not subject to detailed scientific analysis. There has also been an increasing recognition that distinctions must be made between analysis of the risk associated with an event and people's values and preferences. Another important development is the concern by the public that they participate more fully in the decision process on these issues. These concerns were articulated in both the papers and the open discussions at the summer study. The volume does not produce any definitive answers to the dilemmas facing society on how to deal with these new developments. Rather, it raises an important set of questions which need to be considered while at the same time providing a catalog of research needs. These suggested research topics reflect several of the objectives of the current program on risk at the Bundesministerium fuer Forschung und Technologie. In particular, there has been an interest in research on potential costs and benefits associated with new technologies. There was also considerable discussion related to the interaction between technology and society regarding attitudes toward risk, another theme of the current research program. The summer study offered an opportunity for an international exchange of ideas. I hope this volume stimulates future interchanges which address the important issues associated with risk and societal decision making.

Dr. Werner Salz
BMFT
June 1982

Acknowledgement

This book presents a set of papers and discussions from the first IIASA Summer Study on Risk on the theme of "Decision Processes and Institutional Aspects of Risk". The participants in this study represent a cross-section of people from different countries and different disciplines - all of whom have recently been involved in risk research.

The meeting was designed as a forum for discussion on the state-of-the-art of research into the descriptive aspects of risk from an institutional perspective. Its objective was to produce a better understanding of the role that prescriptive techniques such as risk analysis can play in resolving conflicts.

Accordingly, the meeting was structured to encourage an interchange of ideas - the morning sessions being devoted to the more formal presentation of papers with the afternoon sessions reserved for open discussion and small group discussions. We would like to thank the people who chaired these four groups: James Dooley, Oleg Larichev, Dorothy Nelkin, and Paul Schoemaker. The Summer Study was jointly organized by all IIASA researchers involved in risk activities. In this respect, we would like to thank Hermann Atz, John Lathrop, Joanne Linnerooth, Christoph Mandl, Giandomenico Majone, and Michael Thompson for their helpful suggestions regarding the structure of the meeting. We are grateful also to Noel Blackwell for her assistance during all stages of the organization and operation of the meeting. In addition, we would like to thank Valerie Jones and Susan Riley for their contribution to making possible the rapid and accurate preparation of the manuscript. We should like to thank Chris Whipple for providing useful comments and suggestions on all the papers during a visit to IIASA prior to the Summer Study.

We would further like to acknowledge the support of Alec Lee, Area Chairman of Management and Technology, Roger Levien, then Director of IIASA, and C.S. Holling for his continued support of risk activities.

The Summer Study was made possible by financial support from the Bundesministerium fuer Forschung und Technologie of the Federal Republic of Germany. We would particularly like to thank Dr. W. Salz for his helpful assistance and contribution before and during the Summer Study.

Contents

Chapter I

INTRODUCTION

Overview

Howard Kunreuther and Eryl Ley
International Institute for Applied Systems Analysis, Laxenburg, Austria

SETTING THE STAGE

The term risk analysis has generated considerable controversy in recent years as to its meaning and importance with respect to societal decision making. The papers and especially the discussions included in this volume illuminate different aspects of this risk debate. The participants represent diverse disciplinary backgrounds, but all of them have an interest in the decision process and institutional aspects of risk.

The five-day meeting held during the early part of the summer of 1981 was an integral part of a two-year study at IIASA on "Liquefied Energy Gases (LEG): Siting Decisions", supported by the Bundesministerium für Forschung und Technologie of the Federal Republic of Germany. It formed the basis for much of the research and discussions that subsequently took place among researchers at IIASA concerned with problems of risk. We therefore labeled the meeting a summer study to indicate its intended spillover effects. The problems of interest are those which have potential economic benefits to individuals and society, but also have environmental, health, and safety risks associated with them. The siting of large-scale technologies such as energy facilities is a representative example.

Several features of such risk problems can be underscored. First there are many different individuals and groups in society who are affected by a particular decision, each of which has its own goals and objectives, databases and constraints. Each party is also likely to focus on several attributes (e.g., risk to life, environmental risk, costs) and assign different relative importance weights to each of these components.

A second feature of most of these problems is that there is a limited statistical database on which to determine the risk associated with a specific project or activity. Wynne points out that there is a myth of an ideal objective scientific knowledge to deal with policy issues surrounding these problems, while the reality of the situation is that probabilities and losses associated with different events are very ill-specified. As a result, each interested party can interpret data as it sees fit to estimate different risks, and to justify its position.

A third feature of risk problems is the dynamic and sequential nature of the decision process. Different interested parties enter the debate at different stages of the process depending upon the relevant issues being discussed. Allison (1971) points out that at each stage of the process the different parties have different degrees of power and responsibility, as well as conflicting preferences. In order to determine how a particular decision emerges, it is necessary to identify the various issues which were deemed important, to indicate what bargains and compromises were agreed upon, and "to convey some feeling for the confusion".

The sequential nature of the process is also due to the limited time and attention that can be devoted by any subset of the participants to a whole set of different issues and deadlines facing them. As Simon (1967) succinctly put it:

> "...influence over the direction of attention of the political organs is a principal means for effecting action. The notion of power as a tug-of-war between alternatives yields to a notion of power as influence on a sequential decision process, in which actions must be generated as well as chosen and in which attention is a scarce resource" (p108).

These three features of societal problems considered during this Summer Study required us to focus on how institutions and decision processes impact on the treatment of risk. A useful background to this is an understanding of the different meanings attached to the word "analysis". Raiffa in his paper indicates that the US National Academy of Sciences Committee on Risk and Decision Making devoted considerable attention to the role that different types of analysis can play with respect to problems involving risk.

Analysis can simply mean the *assessment* of the risks related to an activity. Vaupel (1981) indicates that the need for this type of analysis arises because various facts uncovered by natural social science research are often insufficient or not well enough established to remove uncertainty. This type of analysis involves the synthesis of disparate and indirectly relevant evidence, both objective and judgmental, in order to assess the probability distribution estimates for the factors of primary concern in the decision problem. Experts with backgrounds in statistics and decision analysis are often called in to undertake these types of assessments. The papers by Mandl and Lathrop and by Lathrop in this volume, focus on this definition of analysis.

Analysis can signify an attempt to understand how the exisiting institutional structure deals with problems involving risk. In this interpretation of analysis one has to specify the different interested parties and their roles, the impact of different legislation and legal constraints as it affects the interaction among these parties, and the different cultural biases and styles between groups or between countries dealing with specific problems. The three papers by Wynne, Ronge, and Thompson and Wildavsky address these issues.

A third descriptive definition of analysis is an understanding of the actual decision making process for dealing with problems where risk is involved. This involves questions on how information is processed by each of the interested parties, the resulting conflicts which emerge in the policy debate, and the way in which final outcomes evolve. The papers by Kasperson and Gray, Dooley and Oseredko, Larichev and Mechitov investigate the decision process in real case study situations.

In a prescriptive context, analysis can mean the types of policies that are proposed for dealing with specific outcomes. In particular, the main theme of the papers and discussions was methods whereby one can provide better information for making decisions, the role of standards and regulations, and also of developing compensation and insurance schemes. The papers by Drake and Long and Fairley concentrate on issues of policy design.

Finally, the word analysis can refer to ways of improving the decision process in order to resolve conflicts. Here one considers mechanisms for resolving conflicts due to disparate evidence (e.g., differences between risk assessments) as well as ways to improve group decision processes. The paper by Kleindorfer and one of the lively interchanges during the discussions were devoted to this definition of analysis.

We will now elaborate on each of these five meanings of analysis, interweaving, where appropriate, the ideas presented in the papers as well as the discussions among the conference participants.

RISK ASSESSMENT

Due to the lack of statistical data and standards that assessments must meet in order to be used as evidence, experts have few guidelines for restricting how they undertake risk assessments. Such broad degrees of freedom have created considerable problems for societal decision making processes.

Oseredko *et al.* indicate that in pipeline siting decisions in the USSR, experts only provide qualitative estimates of many of the parameters and such approximations are influenced by their past experience with a particular feature or attribute.

In their study of a number of different risk assessments, Mandl and Lathrop show how experts use different definitions of risk for the same problem. For example, some estimates of risk focus on a probability of a disaster or accident occurring, and no mention is made of the consequences of the event. The other extreme is that a risk assessment might focus on a maximum credible accident without assigning any probability to its occurrence. Even if all experts were using the same definition of risk, each of them might make a different set of assumptions as to how to estimate the probability or consequence of an event. Hence, two experts focusing on the same situation and using the same definition of risk may arrive at radically different conclusions as to whether it is above or below some specified acceptable level. For these reasons, there was general agreement by the participants that political and social processes greatly influence consideration of risk and that risk should not be looked at in a vacuum.

For example, private consulting firms who undertake risk analyses can have a built-in bias telling the contracting party what they want to hear. Hence the estimates may reflect the different goals and objectives of the relevant interested parties. Linnerooth investigates the implication of several different risk assessments with conflicting estimates concerning the siting process for an LEG terminal in California. She points out that these types of analyses are designed to persuade or support a party argument rather than simply to report results in a statistically meaningful manner.

Finally, we should point out that risk assessments are affected by external events such as accidents. For example, Kasperson and Gray note that following the Three Mile Island accident the Nuclear Regulatory Commission substantially shifted its risk research program toward higher probability/lower consequence events. Similar changes also occurred in the electric power research institutes program in the US.

INSTITUTIONAL ARRANGEMENTS

The focus of attention concerning institutional arrangements was on how analysis is utilized for coping with specific problems in the context of the structure of society. When one looks at risk in this context, it is no longer determined by estimating certain parameters, but rather by a social process. Risk in this sense is viewed as a function of the way different groups in society feel about the problem and interact with each other. In their paper, Thompson and Wildavsky stress the importance of different cultural styles as they affect the group interaction process. Each of the interested groups or parties perceive risk as they want to see it, independent of its physical measurement.

An illustration of how different groups react to the same risk is provided by the analysis of the institutional response to the TMI accident (Kasperson and Gray). The US Nuclear Regulatory Commission was anxious to maintain the *status quo*, despite the fact that the Kemeny Report (the Report of the President's Commission after TMI, 1979) recommended a number of substantial changes in its organizational structure. On the other hand, the electrical utilities who managed the operation of the nuclear power plant used the accident as an opportunity to modify their existing operations. Here two parties analyzed the same situation from their own perspectives and responded to the disaster very differently.

During the meeting considerable attention was devoted to different cultural styles of risk across countries. For example, the USSR attempts to resolve differences between groups through some form of compromise (Oseredko *et al.*), whereas the United Kingdom has more of a hierarchical and consensual style (Wynne), and the US style of societal decision making is more adversarial in nature (Linnerooth). In West Germany, traditional political processes of compromise may have broken down, so that institutions are not functioning in the way they have been in the past (Ronge). Thus, we cannot assume that standard risk analysis techniques that were useful to justify decisions in the past will be able to iron out differences today in West Germany.

DECISION MAKING PROCESSES

This view of analysis stresses the limited ability and desire of decision makers to collect information on which to base their preferences. Interested parties thus attempt to *satisfice* rather than optimize (Simon 1967). For this reason, there is a tendency to use simplistic rules for dealing with particular problems.

This situation is particularly acute after a crisis or accident when many actions have to be taken in a very short space of time (Dooley).

An example of the use of simplified decision rules is provided by Oseredko *et al.* in "The Siting of Pipelines in the USSR", who point out that simple pairwise comparisons were made on separate attributes across alternatives rather than weighting all of the components simultaneously through some type of multi-attribute utility function. One reason for using this type of simplified rule in societal decision making is that the data on which to estimate different attribute values are not very good so that decision makers are hesitant to use sophisticated techniques. In addition, when there are limited alternatives, as in the pipeline case, pairwise comparisons are relatively simple to make. Given that time is an important constraint in these problems, these types of heuristics may be appropriate for the given situation.

Another critical feature of the sequential decision making process is the importance that accidents and random events play. The small statistical database coupled with systematic biases of individuals dealing with uncertainty (Tversky and Kahneman 1974) increase the importance of these exogenous events. Frequently, a disaster jolts the interested parties into action and leads to new regulations or standards to "prevent" future accidents. Kasperson and Gray illustrate this point by detailing new siting regulations which were promulgated after TMI with respect to population density near a nuclear power plant. Drake and Long provide a number of examples of new codes and standards that were triggered by specific accidents and, subsequently used for siting, designing, and operating liquefied natural gas (LNG) and liquefied petroleum gas (LPG) facilities.

In his paper, Wynne points out that human error or incompetent staff are frequently blamed for the cause of a specified accident, whereas in reality, the event may have been due to lack of relevant information and/or faulty organizational procedures. He provides illustrations of these phenomena using examples from the Windscale fire of 1957 and the TMI accident.

A pervasive problem regarding the societal decision process is the conflict among the different interested parties. The analyses utilized by each of the parties to defend their objectives are generally not based on tested assumptions, nor are they easily challenged in hearings or public enquiries. The resolution of these conflicts will depend on the nature of the problem, the degree and timing of public participation, as well as the relevant style of risk management.

Recently, the Risk Group at IIASA has developed a framework of the decision making process for problems such as the siting of facilities based on new technologies. This multi-attribute, multi-party (MAMP) approach considers the role of many interested parties in this specific concern (Kunreuther *et al.* 1982). It emphasizes the potential for conflict emerging among the interested parties as a result of their different objectives, mandates, and information sources. Linnerooth illustrates the use of the MAMP framework in the context of the siting decision of an LNG terminal in California.

POLICY DESIGN

Market-based solutions for dealing with risk are of limited usefulness in dealing with the types of societal decision making problems covered in this summer study. One reason is that the lack of a good statistical database to estimate risk makes it difficult to determine the expected costs and benefits associated with different programs.

Kleindorfer suggests three additional reasons why such policies will not work well. The first relates to the fact that these problems are classified under the heading of public goods or public bads where individuals are involuntarily exposed to certain risks. A second factor that we have already alluded to is the information collection and processing difficulties facing all interested parties. Thirdly, at the level of the firm, the catastrophic loss potential associated with particular risks makes it difficult for enterprises to obtain protective measures such as insurance.

For these reasons, social institutions for dealing with these types of problems have emerged. Therefore a critical factor is a thorough understanding of the political and social processes associated with particular problems. We will now briefly review the policies which were proposed in the papers and discussions during our meeting. The potential success of any of these proposals depends on the problem context, the existing institutions and decision processes of the relevant interested parties. We will provide examples of where certain programs have worked well in specific countries.

Providing better information. One way that analysis can help to improve decision making is to provide better information to all the parties who are affected by particular programs. One very positive example of this is in Sweden, where the government financed a program to inform the public about energy and nuclear power. The program involved 8000 study circles each consisting of approximately ten members who came together to discuss those energy-related questions they felt to be most important. The expectation on the part of the Swedish government was that this type of interchange would create a more favorable attitude toward government policy. However, the reports from these groups suggested that there was a continued uncertainty and ambivalence which resulted in lack of consensus on the issue of nuclear power (Nelkin and Pollack 1979).

Another way of providing information to different parties is through certain types of decision aids which can be used to help determine the relative merits of different alternatives (e.g., whether to site a facility at location A, B, C, or not at all). Lathrop feels that one can broaden traditional risk analysis from looking only at an expected number of fatalities, to one covering a broader set of attributes (e.g., expected health effects, catastrophic potential and unequity of impacts). He contends that this information can be incorporated through an expanded multi-attribute utility model so that the decision maker has a better appreciation of the social and political factors that may impact on final outcomes. Although this approach may be useful for a single decision maker, it still does not address the problems of conflict resolution when the different parties have not been able to reach a consensus.

In their papers, Dooley and Kleindorfer both recommend computer-aided decision support systems which will enable policy makers to determine the impact of different policies on physical phenomena under a set of scenarios that they create. For example, assume the policy maker was interested in the consequences of a fire to an LNG storage tank which was proposed for a given location. Based on models of vapor cloud dispersion he could investigate the potential destruction and fatalities under different assumptions with respect to the way the wind blows. Furthermore, he could place the LNG terminal at different points within the community or at different sites to investigate the potential catastrophic impact under different wind conditions. This type of interactive analysis enables different interested parties to see the effect of alternative assumptions and physical models (e.g., vapor cloud dispersion models) on outcomes.

Developing standards and regulations. Another class of policies relates to specific codes and standards which make structures safer or which impose restrictions on where projects can be located. These types of regulations serve many different interested parties simultaneously. Residents, for example, may feel safer if a facility is sufficiently far away from their homes. At the same time, the government may wish to avoid the higher prices of a potentially catastrophic accident in a densely populated area. As Drake and Long point out, industry is also frequently in favor of certain standards and regulations because it prevents the less responsible firms from damaging the industry's reputation. One of the open questions with respect to standards and regulations is how to estimate the potential costs and benefits of these measures to the different interested parties. In addition, there is then a need for a regulatory body to monitor, control, and enforce these measures.

Instituting insurance policies and compensation measures. The principal purpose of these programs would be to force those parties (e.g., industry, government) who stand to benefit from particular programs also to bear an appropriate share of the costs should there be damage to the environment or losses to property and life.

Insurance programs and compensation schemes directly address the distribution of gains and losses from specific programs. Such distribution questions are critically important in political decision processes. As Robert Behn (1981) argues:

> "In contrast to the analyst, the politican is most interested in distribution. Political influence is exercised not by individual citizens but through the intermediation of organized constituencies, as an elected official or political appointee of one, the policy politician is responsive to constituencies—not to the analysts' abstract notion of aggregate public welfare. The politician is deeply concerned about how much his constituents will benefit and how much they will pay."

One example of an institutional response in this direction is the self-regulation measures adopted by the public utilities following the TMI accident. Kasperson and Gray point out that the utilities have cooperated to create an insurance pool to protect themselves against the potential extraordinary cost which would accrue to any utility following a future major nuclear accident. Fairley recommends that producers who benefit from specific projects be made liable for all the costs they impose upon others. He points out that this can take the form of self-insurance, commercial insurance, industry risk pooling (as in the TMI example) and/or government-managed compensation funds. The previous history of societal risk bearing in most countries suggests that these schemes would be difficult to implement if there are catastrophic risks which tax the reserves of private industry. The Fairley proposal does suggest the importance of allocating responsibility prior to a disaster rather than determining who will pay after the event occurs.

Much of the opposition to the proposed high-risk projects comes from relatively small groups of individuals, often residents of the area, who feel that they would suffer losses in property values and would have to bear the costs of health and environmental risks. O'Hare (1980) has characterized this general problem as "not in my backyard". He has proposed a particular type of compensation scheme whereby each community considered as a potential site, determines a minimum level of per capita compensation, and it would be willing to make a legal commitment to have the project in their backyard if the compensation is paid. The applicant would include such compensation as relevant in calculating the siting costs associated with locating the facilities in community A, B, or C. The final decision would then be made by the applicant taking into account the amount of compensation he would have to pay residents in each of these three localities.

Whether or not such a compensation scheme is a useful policy description depends on the specifics of the situation and the cultural differences in risk management style. In this connection, it would be interesting to ask what types of payments would have been required to please the citizens of Oxnard, California, so that an LNG terminal could have been located there. What would the Sierra Club require in payments for them to support a site which might have adverse environmental effects? As indicated in the papers and discussions at the meeting, these questions can only be answered in a real world problem context. They do reflect an increasing concern of social scientists and policy analysts in dealing with windfalls or wipe-outs from specific accidents which involved the public sector. Kunreuther and Lathrop (1982) point out that the final outcome is likely to represent a balance between the political restraints and economic criteria. They refer to a quote by Wildavsky (1981) which states:

> "The criterion of choice in politics and markets is not being right or correct as in solving a puzzle, but agreement based on interaction among partially-opposed interests" (p131).

DESIGNING ANALYSES FOR IMPROVING PROCESS

Wynne points out in his paper that there is a myth that consensus between different interested parties is possible because there is only one objective truth.

He contends that the reality is quite different since there is no single scientific truth. If consensus occurs, it is due to default rather than agreement. Much of the discussion at the meeting revolved around the question "How do we improve the decision process given that there is not a single scientific truth? We will now briefly summarize the major points coming out of the papers and the discussion.

Establishing Credibility of Analyses

In general, existing institutions in most countries are inadequate to deal with problems of conflicting evidence and polarized expert opinion with respect to questions such as risk assessments. One of the recommendations of the CORADM Report is that one should report honestly on the basis for disagreements among experts in the assessment process and then have policy makers determine what to do with the data in their evaluation process (Raiffa). The committee pointed out that experts will disagree due to their different experiences, their different paradigms, and the different types of information they utilize.

One way of dealing with this problem is to establish rules of evidence where one can better understand the differences in experts' risk analyses. Lathrop and Linnerooth (in press) suggest a set of guidelines establishing these rules of evidence. In particular, they stress the importance of defining the risk being assessed, clarifying the assumptions and error bounds, and indicating the conditional nature of specific analyses which were undertaken.

A principal problem which currently exists in implementing these guidelines is the lack of an institutional mechanism for reviewing different risk assessments. Ackerman *et al.* (1974) point out that the traditional approaches such as legal responses, agency hearings and judicial reviews have inherent limitations with respect to evaluating these conflicting assessments. Such evaluations are especially difficult for classes of problems where there are no statistically based measures of risk. These authors propose the establishment of a review board to examine different assessments. Under their proposed procedure members, all of whom would be trained in subjects fundamental to technical analysis, would provide a written report evaluating the impact of specific assessments for specific issues (e.g., population risk, environmental impact). Particular attention should be given to identify the empirical basis of the set of findings and to determine how well the analysis is grounded in scientific theory.

Group Procedures for Facilitating Collective Action

When different interested parties disagree on a preferred set of alternatives for specific projects, there is a need for each to understand the other side's decision. Kleindorfer points out that this is particularly important when there are conflicts of values (e.g., the importance of safety or environmental impact on a project) or conflicts of fact (e.g., different risk assessments). In his paper he outlines a number of approaches for dealing with this problem. Particular attention is given to strategic planning models when one tries to mesh an ideal state with the *status quo* through scenario generation.

It may also be possible to involve the different participants in role-playing activities. Holling and his colleagues at the Institute of Resource Ecology have conducted workshops on role-playing where each of the participants specifically takes on the role of one of the relevant actors in a societal problem. Prior to the workshop, each participant interviews a particular individual whom he is representing so that he can understand his objective and relevant information being considered in evaluating different alternatives (see Holling 1981).

Institutions for Resolving Conflicts

There was considerable interchange in the discussion on different proposed review mechanisms for helping to resolve conflicts among different interested parties. There was general consensus among the group that the type of procedure is critically dependent on the specific problem being considered and the social and cultural environment. For example, reference was made to the Canadian experiment in which the Berger Commission was created to assess the Mackenzie Valley Pipeline controversy. As part of this project, intervener groups were given financial support to develop their case. The Dutch Science Shops were also cited as an example of a public enquiry which evaluates the reaction of the interested parties who are affected by projects that impact on the environment.

These examples are elaborated upon in the discussions which appear in Chapter 5. Finally, we should point out that to understand risk decision making one has to consider the cultural milieu. Thompson and Wildavsky in their paper put forward a theory as to why different types of people in the same situation react differently. One of the stimulating features of the Summer Study was the different reactions of the participants (from seven different countries both East and West) when we discussed various issues.

ORGANIZATION OF THE PAPERS

The first paper by Linnerooth outlines a descriptive framework of the decision making process for problems such as siting facilities. The approach, developed here at IIASA is a multi-attribute multi-party (MAMP) framework of choice in a dynamic setting. It considers the potential for conflict emerging between interested parties as a result of their differing objectives, mandates, and information sources. On the prescriptive side the approach explores the roles that analyses, including risk analysis, and decision theory can play in providing a more systematic basis for making decisions. The framework is illustrated in terms of a specific problem, that of the LNG siting decision in California, USA.

Raiffa, in his paper, describes the work of the US National Academy of Sciences Committee on Risk and Decision Making (CORADM). The objectives of this committee were very similar to those of our Summer Study. Much of the attention of the committee was spent on the role of analysis in dealing with problems which have adverse effects on health, safety, and the environment. In particular, the paper addresses the difficulty of decomposing risk assessment from risk evaluation, a theme which is stressed in other papers as well as in the dicussions.

Chapter 2 deals with "Risk Assessment in a Problem Context". The first paper by Mandl and Lathrop examines the assessment of risk to life in catastrophic accidents due to a large-scale technology (LNG) facility. Two primary goals are to present and compare various procedures of risk assessment and to quantify and compare risks at four facilities. The paper concludes that there are disagreements between the experts as to how to quantify risk, which models to use, and what to include and exclude in risk assessment. This is due to new technology (limited data), new techniques of risk assessment for LEG terminals (new approach), as well as the different needs of the client.

Kasperson and Gray, in their paper, report on the response and changes which occurred after TMI and the impact of the Kemeny Report on nuclear safety. Each of the interested parties reacted to the accident and report in different ways, reflecting their own goals and objectives. The authors point out that the media had considerable coverage of the Kemeny Report prior to its being issued as well as immediately thereafter. However, they exhibited lack of sustained analysis and the subsequent societal response. Industry undertook a number of reforms in managing nuclear power plant safety even before the recommendations of the Kemeny Commission. In contrast, the Nuclear Regulatory Commission's response to both TMI and the Kemeny Report has "been more delayed and uneven". The paper concludes that the long-term response to TMI and the report may be to further self-regulation in nuclear power.

Dooley's paper was stimulated by the organizational and institutional response to the Missassauga, Ontario, accident where 24 cars of a train derailed, 19 of which carried hazardous material. It looks at the characteristics of crises and then suggests ways of improving decision making. In particular, Dooley points out that decision making in a crisis seems to be described by Simon's procedural rationality model in which individuals have limited time to deal with the situation and sparse information. Decision makers will take actions based on past experience in similar situations.

Oseredko *et al.* describe the site selection process for locating gas pipelines in the USSR where there are many different attributes to be considered. Each of the interested parties in the debate considers a subset of these attributes so that at the outset there is no consensus on which one of the several alternatives to choose. The paper illustrates the sequential nature of the decision process, the use of heuristics for arriving at a final state through compromise, and the limited role that experts play in arriving at a final solution.

The final paper in Chapter 2 by Drake and Long, discusses the development of codes and standards used for siting, designing, and operating hazardous facilities such as LNG and LPG plants. Drake indicates the importance of unexpected accidents in triggering regulations and points out that standards will continue to evolve as accidents indicate weaknesses in present requirements.

Chapter 3 focuses on "Institutional Aspects of Risk". Ronge, in his paper, indicates that the traditional political processes of compromise in bargaining have broken down and hence one has to re-evaluate the role that analysis plays in the societal decision making process. In particular, the author focuses on West Germany and points out that parties that lose in the policy debate frequently resort to violence. He offers as reasons for these changes the ecological movement, the "small is beautiful" philosophy leading to opposition of large technologies, and the general antipathy to social institutions. The paper concludes that risk research needs a framework grounded on theoretical principles which incorporates political processes.

The following paper, by Wynne, is concerned with myths regarding the operation of institutions and the use of scientific rationality in risk assessment. Society is operating under the myth that there is objective scientific knowledge which can be used to deal with policy issues associated with risk. The reality is we cannot achieve this ideal. By striving to reach it we may create more problems than we solve. The paper argues that there is a need to study how scientists, organizations and institutions function in the context of the political process so that we can replace the set of current myths with a more realistic picture of the world.

Chapter 3 ends with "A Proposal to Create a Cultural Theory of Risk", by Thompson and Wildavsky, who attempt to answer the question as to why different types of people facing a similar societal problem react in different ways. A cultural theory of risk is developed which characterizes five different groups on the basis of the intersection of two dimensions: groups (the extent to which an individual is involved in social groups) and grids (the extent to which an individual is involved in hierarchical arrangements). Each of the five resulting cultural categories has its own rationality which is likely to be contradictory. Risk debates are likely to differ between cultures because of different constellations of groups in each culture. On the basis of this theory, one can give policy makers insights into why there are profound disagreements over risk, and under what situations one can hope to resolve conflicts between parties.

Chapter 4 is concerned with "Decision Processes and Prescriptive Aspects of Risk". The principal theme of Lathrop's paper is that there are two perspectives on risk: the technical and psychological, each of which looks at the societal and decision making problem in different ways. The technical perspective focuses on a single index, the expected number of fatalities from a given alternative. The psychological perspective looks at a variety of different factors in evaluating the risk. At the prescriptive level, the paper proposes an expansion of a multiattribute utility function so that it recognizes political realities in social concerns which affect risk management.

In his paper, Kleindorfer indicates the information and process difficulties associated with market-based solution and regulatory mechanisms for solving societal decision making problems. The key challenge in risk management is to find ways of reconciling conflicts between the parties because of the different assumptions, goals, and objectives, as well as the knowledge base on which each group operates. The paper proposes the use of group processes for reconcilling differences building on the work of Ackoff *et al.* on strategic planning. The approach relies on scenario generation towards an ideal state which requires stating the necessary assumptions which each group has made to achieve this ideal state. By coupling this approach with specific methods (e.g., Saaty's analytic hierarchy approach, computer-based decision support systems, role playing) one may be able to get the decision maker to see differences in the assumptions. In this way interested parties may be able to achieve consensus through compromise solutions.

In the concluding paper, Fairley argues for market-based solutions to problems of societal risks so that risk producers are liable for the costs they impose on others. By shifting the responsibility for catastrophic accidents from society to producers, the author feels that risk producers will undertake thorough risk assessments since they have a stake in the accuracy of the data. In this way risk assessments will focus on a number of other attributes besides loss of life, such as injury and property damage.

REFERENCES

Ackerman, B., Rosa-Ackerman, S., Sawyer, J., and Henderson, D. (1974) The Uncertain Search for Environmental Quality. New York: The Free Press.

Allison, G.T. (1971) Essence of Decision. Boston: Little, Brown and Company.

Behn, R.D. (1981) Policy analysts and policy politicians. Policy Analysis 6(2): 21-46.

Holling, C.S. (1981) Science for Public Policy: Highlights of Adaptive Environmental Assessment and Management. R-23. University of British Columbia, Vancouver, Canada.

Kunreuther, H., and Lathrop, J. (1982) Hazardous facilities: siting lessons from LNG. Risk Analysis (in press).

Kunreuther, H., Lathrop, J., and Linnerooth, J. (1982) A Descriptive Model of Choice for Siting Facilities. WP-82-18. Laxenburg, Austria: International Institute for Applied Systems Analysis.

Lathrop, J., and Linnerooth, J. (1982) The role of risk assessment in a political decision process, in: P. Humphreys and A. Vari (Eds.) Analyzing and Aiding Decision Processes. Amsterdam: North-Holland.

Nelkin, D., and Pollack, M. (1979) Public participation in technological decisions: reality or grand illusion. Technology Review, September pp55-64.

O'Hare, M. (1977) Not on my block you don't: Facility siting and the strategic importance of compensation. Public Policy 25:409-58.

Simon, H. (1967) The changing theory and changing practice of public administration, in: I. de Sola Pool (Ed.) Contemporary Political Science: Toward Empirical Theory. New York: McGraw-Hill.

Tversky, A., and Kahneman, D. (1974) Judgment under uncertainty: heuristics and biases. Science 185:1124-31.

Vaupel, J.W. (1981) Truth and consequences: some roles for scientists and analysts in environmental decision-making, in: W.A. Magat (Ed.) Improving Environmental Regulation. Cambridge, Mass.: Ballinger.

Wildavsky, A. (1981) Rationality in writing: Linear and curvilinear. Journal of Public Policy 1:25-40.

Siting an LNG Terminal in California: A Descriptive Framework

Joanne Linnerooth

International Institute for Applied Systems Analysis, Laxenburg, Austria

1. BACKGROUND

Liquefied natural gas (LNG) is, as the name suggests, a gas that is liquefied for purposes of transportation. To liquefy natural gas, its temperature is reduced to -160^{0}C, at which point the volume of the gas is reduced to approximately one-six hundredth of its original volume. Presently, there are 16 receiving plants in the world, primarily in Japan, in Western Europe, and in the US. A typical tank has a volume of 80 000m^3; there are usually from two to four of these tanks at a receiving terminal. The early ships had a capacity of something around 27 000m^3, present ships as high as 130 000m^3. It was estimated that in 1981 there would be at least 57 LNG carriers operating in the world with a combined capacity of over 5.21 million m^3.

The capital costs of a typical operation include the costs of the export terminal, the receiving terminal, and the ships. The cost of a typical export terminal is around $750 million; a receiving terminal with four tanks costs something in the order of $150 million; and three methane tankers, a minimal number, add another $450 million. This totals more than a billion dollars (Office of Technology Assessment 1977). The throughput of a typical plant, as calculated from Point Conception (see Mandl and Lathrop 1981), is approximately 15 000 megawatts (MW) of electricity. This is approximately equivalent to the throughput of 15 nuclear power plants. In sum, the import or export of LNG is a highly capital-intensive and energy-intensive operation.

In the event of a ship or terminal accident, a significant amount of LNG could be spilled, which would "boil off" into a methane cloud. Since the dispersion characteristics of methane clouds are poorly understood there is a great deal of uncertainty involved in predicting accident consequences. Yet, the present state of knowledge indicates that at some very low probability an LNG accident could result in a cloud covering several miles before igniting. Depending on the population density of the area covered by the cloud, the possibility exists, albeit at a low probability, for a catastrophic accident.

A liquefied energy gas terminal, therefore, promises to yield benefits to society, but only at a cost of potential catastrophic accidents. The siting of these large-scale facilities presents a formidable challenge to political risk management processes. There are two features of these problems that make them particularly difficult to resolve. First, the gas consumers who benefit from the terminal do not always bear the risks, which fall on a small group of people living in the vicinity of the terminal. The problem is how to distribute the costs and benefits. Hence, there is the potential for conflict among the interested parties. A second feature of the siting problem is the absence of a database that provides conclusive statistical evidence on the likely performance of the new

---

*The research reported in this paper is supported by the Bundesministerium für Forschung und Technologie, FRG, contract No. 321/7591/RGB 8001. While support for this work is gratefully acknowledged, the views expressed are the author's own and are not necessarily shared by the sponsor.

technology and the probability distribution associated with potential accidents. Each of the interested parties may thus provide different estimates of the chances and consequences of certain events. There are *no* objective measures to settle these differences.

The IIASA Risk Group is investigating how decisions with these features are taken. For this purpose, the siting of LEG terminals in four countries—the FRG, the Netherlands, the UK, and the USA—have been selected as case studies. The approach taken at IIASA builds on the notion that a good understanding of how societies make siting decisions, i.e., the institutions involved and the uses made of scientific expertise, is necessary to improve upon these processes.

2. THE DESCRIPTIVE FRAMEWORK

In this section, I will briefly discuss the framework used to compare the siting process in the four countries under study, and will illustrate this framework in the context of the siting process in the United States. The framework emphasizes the involvement of many interested parties and the diverse concerns, or attributes, of these parties. Hence we have labeled it the multi-attribute, multi-party (MAMP) approach. In developing this structure, we have been greatly influenced by the concepts discussed by Braybrooke (1974) where he looks upon the political system as a machine or collection of machines for processing issues.

2.1. Rounds

The decision process can be separated into different rounds which we label by capital letters, A, B,... A round is simply a convenient device to illustrate a change in the focus of discussions. This new focus or direction can be triggered by (1) a key decision taken (or a stalemate reached due to conflicts among parties), or (2) a change in the context of the discussions due to an unanticipated event, the entrance of a new party, or new evidence brought to the debate. Rounds are simply a convenient way of segmenting the decision process; they are not unique and can be simultaneous or overlapping.

The problem in each round is defined by a set of issues, decision constraints and procedures. Braybrooke (1974) refers to an "issue-circumscribing" phase where the alternatives for discussion are bounded by generally accepted, though not necessarily irrefutable, facts and values, e.g., "it is technically feasible to import LNG to California" or "California needs LNG". As we have already noted the problem is also formulated or defined by the decisions from earlier rounds. Clearly the problem-defining process will have an effect on final outcomes through this latter type of bounding constraint.

2.2. Problem Formulation

A round of more or less official discussions is initiated by a formal or informal request. Informal discussions may be initiated simply by such actions as a request for information on the part of one of the parties or a request for preliminary discussions. Because the particular form of how the round is initiated may further define or limit the bounds of the discussion, the careful scrutiny of their wording is important. For example, it may make a difference in the decision process if the question is framed as "is there a site which is appropriate?" or "which of the sites x, y, and z is appropriate?" Braybrooke (1974) refers to the first question as a "which-question" and the second as a "whether-question". Which-questions demand more complicated considerations and detailed thinking, while whether questions can be approached with simpler rules of thumb and heuristics.

2.3. Interaction

To understand a particular pattern of institutional choice it is necessary to analyze a set of policy actors, their interactions with one another at different stages of the process, and the information available to them. A party's evaluation of an alternative is based on its estimation of the levels and values of each attribute resulting from that option, the value of each of those levels, and the relative importance given to each attribute. Another party might have different estimates of the effects of an option, different costs and benefits resulting from those effects, or may assign a different relative importance to each of the attributes. Because of any of these differences one party may rank alternatives differently from another.

The interaction among the parties is represented by the *main arguments* each brings to the debate in support of or in rejection of each of the alternatives at hand. Those arguments may relate to only one or two attributes. It is *not* suggested here that the arguments presented for or against a particular proposal necessarily reflect a concern of the party making the argument. For example, a party opposed to a site because of its concern for environmental quality may present an argument using seismic risk as the main reason to reject the site. The argument may be selected to maximize the effectiveness of the argument, not to reflect the actual concern of the party. The argument reflects a *strategy* on the part of the actor in support of or opposition to the proposal. The strategy of the actors can reveal a number of underlying motives and desires of those concerned and may be essential in understanding the interpretation and use of scientific evidence, including risk analyses.

2.4. Concluding a Round

The round is concluded by a decision, a stalemate, a change in information (changing the focus of the debate and hence initiating a new round), or an unanticipated event aborting the discussions and requiring a new round of inquiry. Each decision can, in turn, be described by the trade-offs implicit in the choice made. These trade-offs may not be explicitly recognized by the decision maker, or not explicitly analyzed in the process of making the decision.*

3. APPLYING MAMP: THE SITING OF THE CALIFORNIA LNG TERMINAL

In the late 1960s, faced with projections of decreasing natural gas supplies and increasing need, several California gas utilities began to seek additional supplies. In 1974, Western LNG Terminal Company (Western), which was formed to represent the LNG interests of the gas utilities, applied for approval of three LNG import sites on the California coast: Point Conception, located on a remote and attractive part of the coast; Oxnard, a port city; and Los Angeles, a large harbor metropolis. The LNG would be shipped from southern Alaska, Alaska's North Slope, and Indonesia. At the time of writing, Point Conception, the one site remaining under active consideration, is still pending approval. This section describes the interested parties, procedures, decisions, and events of this lengthy process (for a more complete review see Linnerooth 1980, Lathrop 1980, Kunreuther *et al.* 1981).

3.1. Interested Parties and Relevant Attributes

To structure the siting process we need to have a good understanding of the different concerns of the interested parties. In the California case, there were

---

*The distinction between a "decision maker" and a decision resulting from a process is an important one since the person responsible for the decision often cannot be identified (see Allison 1971, Majone 1979).

Table 1. Principal party-by-attribute matrix for LNG siting in California.

Attributes		Principal parties*							
		Applicant	Government					Interest groups	
			Federal	State			Local		
		Utilities P_1	FERC P_2	CCC P_3	CPUC P_4	Legis. P_5	municip. P_6	Sierra Club P_7	Local citizens P_8
RISK									
Supply interruption	X_1	•	•		•	•			
Population	X_2		•	•	•	•	•	•	•
Earthquake	X_3		•	•	•	•			
ENVIRONMENTAL									
Air quality	X_4		•		•	•		•	
Land use	X_5		•	•	•	•	•	•	•
ECONOMIC									
Profit consideration	X_6	•							
Price of gas	X_7		•		•	•			
Local economic benefits	X_8						•		

*Key to party acronyms, abbreviations:

FERC: Federal Energy Regulatory Commission, or in the first two rounds of the process, its preceding agency, the Federal Power Commission.
CCC: California Coastal Commission.
CPUC: California Public Utilities Commission.
Legis: California State Legislature
Municip: Municipal government.

Source: Kunreuther *et al.* (1981).

three categories of concern which are relevant: risk aspects, economic aspects, and environmental aspects. Each of these concerns can be described by a set of attributes. Table 1 depicts an interested party/attribute matrix showing the main concerns of each of the relevant groups over this seven-year period.

The attributes listed have been selected to reflect the nature of debates in the process, that is, to reflect the attributes as perceived by the parties in the debate, rather than to characterize in some logical analytical manner the alternatives. For example, population risk (X_2) involves the risk to life and limb to neighbors of the LNG terminal due to accidents including those induced by earthquakes. Earthquake risk (X_3) which involves both population risk and supply interruption risk due to earthquakes, is included as a separate attribute since it was handled as such in the process.

The filled cells in Table 1 indicate which parties pay particular attention to which attributes. Naturally, many of the parties care about all the attributes listed. However, either because of the incentives directly felt by the party or because of the role the party plays in society, each party makes its decisions based upon a particular subset of the attributes. For example, while the applicant is certainly concerned with environmental quality and risks to the population, its primary responsibilities and concerns are earning profits for shareholders and delivering gas reliably to consumers. Its actions are apt to be motivated by concerns for profits and gas supplies, and constrained by political and legal limits set by other parties' concerns for safety and the environment. Likewise, the Sierra Club cares about reliable gas supplies, but receives membership dues for being primarily concerned with environmental quality. Consequently, in a situation where a proposed action increases the reliability of gas supply at the expense of environmental quality it is reasonable for the applicant to favor the proposal and the Sierra Club to oppose it. These differences in primary concerns may determine a great deal of the behavior of the political decision process, and explain how that process is apt to differ from the single decision maker postulated by normative evaluation approaches. The important message of Table 1 does not lie in the details of exactly which cells are filled, but lies in the generally great differences in the columns of the table. That is, the different parties in the process care about different subsets of the attributes.

The applicant, Western LNG Terminal Associates, was a special company set up to represent the LNG siting interests of three gas distribution utilities: Southern California Gas Company, Pacific Gas and Electric, and El Paso Natural Gas Company. As domestic gas supplies seemed to be diminishing in the late 1960s, the gas utilities perceived an increased risk of supply interruption, which could be mitigated by additional supplies such as LNG. Quite naturally, the applicant was primarily concerned with profitability (X_6) and secure supplies of gas (X_1).

At various government levels there are five principal parties. The Federal Energy Regulatory Commission (FERC) in the Department of Energy is the principal body at the federal level which determines whether a proposed LNG project is in the public interest and should be allowed. In making its judgment it considers primarily the following attributes: risk factors (X_1, X_2, and X_3), environmental guidelines as reflected in air quality (X_4) and use of land (X_5), and the expected LNG price (X_7).

At the state level, the California Coastal Commission (CCC) was created in 1976, and has the responsibility for the protection of the California coastline. Its primary concerns with respect to LNG siting are with the use of land (X_5) and the associated risks (X_2 and X_3) from building a terminal at a specific site. The California Public Utilities Commission (CPUC) is the principal state body involved in power plant issues and is primarily concerned with the rate-setting process. Hence, it focused on the provision of energy to California residents and the need for gas (X_1) and the proposed price of the product (X_7). In addition, it has responsibility for evaluating the impact that a proposed facility would have on the environment and safety. The California state legislature is ultimately responsible for the outcomes of any siting process. It determines which state and local agencies have final authority to rule on the feasibility of a proposed site. In addition, it can set standards to constrain any siting process. Hence the concerns of the legislators range over economic, environmental and safety attributes, as shown in Table 1.

At the local level, the city councils evaluate the benefits of a proposed terminal in their jurisdiction in terms of the tax, business revenues, and jobs (X_8) it promises to provide. The councils try to balance this positive feature with the impact that the facility would have on land use (X_5) and risk to the population (X_2). Finally, the public interest groups, represented by the Sierra Club and local citizens' groups, are primarily concerned with environmental and safety issues.

3.2. The Decision Process

The siting process in California (which is not yet terminated) can be characterized by four rounds of discussions as shown in Table 2, which provides a summary of how the problem was defined, the initiating event, and how the discussions were concluded. The remainder of this subsection discusses in more detail the decision process within each of the rounds. The main elements of round's A, B, C, and D are described in Tables 3, 4, 5, and 6 respectively.

Table 2. Summary of rounds in California LNG siting case.

		DATE
ROUND A		
Problem formulation:	Should the proposed sites be approved? i.e., Does California need LNG, and if so, which, if any, of the proposed sites is appropriate?	
Initiating event:	Applicant files for approval of three sites.	September 1974 (34 months)
Conclusion:	Applicant perceives that no site is approvable without long delay	July 1977
ROUND B		
Problem formulation:	How should need for LNG be determined? If need is established, how should an LNG facility be sited?	
Initiating event:	Applicant and others put pressure on state legislature to facilitate LNG siting.	July 1977 (2 months)
Conclusion:	New siting process set up that essentially assumes a need for LNG, and is designed to accelerate LNG terminal siting.	September 1977
ROUND C		
Problem formulation:	Which site should be approved?	
Initiating event:	Applicant files for approval of Point Conception site.	October 1977 (10 months)
Conclusion:	Site approved conditional on consideration of additional seismic risk data.	July 1978
ROUND D		
Problem formulation:	Is Point Conception seismically safe?	
Initiating event:	Regulatory agencies set up procedures to consider additional seismic risk data.	
Conclusion:	(Round still in progress).	

Round A began in September 1974, when the applicant filed for approval of three sites on the California Coast—Point Conception, Oxnard, and Los Angeles—to receive gas from Indonesia. The application raised two central questions which defined the problem addressed in Round A: Does California need LNG, and if so, which, if any, of the proposed sites is appropriate?

The agenda for discussion was more narrowly defined at this stage. The wheels of the process were set into motion, not by a broadly based energy policy question initiated in Washington, but by a proposal from industry for three preselected sites. The importance of this process—where the initiative was taken *first* by industry—in preselecting the agenda for debate cannot be overemphasized. The initiating proposal framed the problem as "Should the proposed LNG sites be approved?" and *not* "Should California have an LNG terminal in view of the alternatives, costs, risks, etc?" Setting the agenda in this manner did not preclude the "need" question from entering the debate, but it did ensure that the question was only considered in the context of a siting application.

Table 3 also specifies the relevant interested parties who were involved in the interaction phase of round A. Those parties which had formal decision power are marked with an asterisk. There were four primary attributes which were utilized in the ensuing debate among the parties. The need for LNG or the risk of an interruption in the supply of natural gas (X_1) supported the locating of a terminal in at least one of the three proposed sites. While environmental land use considerations (X_5) suggested a non-remote site (Los Angeles and Oxnard), the risks to the population (X_2) argued for siting the terminal in a remote area (Point Conception). Finally, concerns about earthquake risk brought about opposition to the Los Angeles site, which was found to be crossed by a significant fault.

Table 3. Elements of round A.

Problem formulation:	Should the proposed sites be approved? i.e., Does California need LNG, and, if so, which, if any, of the proposed sites is appropriate?					
Initiating event:	Applicant files for approval of three sites.					
Alternatives:	Point Conception:	A^1				
	Oxnard:	A^2				
	Los Angeles:	A^3				
	Any combination of:	A^1, A^2, A^3				
Interaction:	Involved parties		Attributes used as arguments			
	Applicant	P_1	X_1			
	FERC*	P_2	X_1		X_3	
	CCC*	P_3		X_2		
	City councils*	P_6		X_2		X_5
	Sierra Club	P_7		X_2		X_5
	Local citizens	P_8		X_2		X_5
Key decisions:	(1) CCC concerns over population risk implies that A^1 is preferred over the other two sites. (2) FERC would not approve A^2 because the seismic risk is greater than a prescribed acceptable level.					
Conclusion:	Applicant perceives a stalemate, i.e., that no site is approvable without long delay.					

*Interested party with responsibility for decision(s).

The interaction phase of round A (see Table 2) indicates the attributes used as arguments by each of the major involved parties. It is important to distinguish this listing of attributes from that in Table 1. While Table 1 specifies which attributes are of *primary concern* to each party, Table 3 specifies which attributes were *used as arguments* by each party. Thus while the applicant is concerned with both profit considerations and supply interruption risk, its arguments in support of each site stressed supply interruption risk.

Two key decisions were made during round A. First, the CCC, concerned about the catastrophic potential of LNG, implied that they were likely to favor Point Conception over the non-remote sites due to concerns over population risk. Specifically, the CCC advised Western to pursue at least one site in a remote area since they would deny approval to any non-remote site which was not considered safe. Second, the FERC indicated disapproval of the Port of Los Angeles as an acceptable site because a recently discovered earthquake fault increased the seismic risk above a prescribed acceptable level.

The round was concluded with a possible stalemate, at least as perceived by industry (Ahern 1980). Los Angeles would not receive federal (FERC) approval, Oxnard might not receive state (CCC) approval, and Point Conception would face difficult approval challenges at the county and state (CCC) levels because of its adverse land-use impacts.

The stalemate of round A formulated the problem for round B. It was clear to all the parties involved that it was difficult, if not impossible, for the applicant to gain approval for a site under the existing siting procedure in California. In particular, there were possibilities of vetoing proposals at either the federal, state, or local level, as evidenced by the respective reactions to the three proposed sites. Rather than trying to operate within the existing constraints of the process, the interested parties in the process frequently tried to change the rules of the game (Majone 1979).

Table 4. Elements of round B.

Problem formulation:	How should need for LNG be determined? If need is established how should an LNG facility be sited?
Initiating event:	Applicant and others put pressure on state legislature to facilitate LNG siting.
Alternatives:	Consider offshore sites: B^1 Consider remote onshore sites: B^2 Consider non-remote onshore sites: B^3 One-stop licensing B^4 Licensing agency: CPUC = B^5, CCC = B^6, CEC = B^7* Any consistent combination of B^1 through B^7.

Interaction:

Involved parties		Attributes used for arguments		
Applicant	P_1	X_1		
CCC	P_3		X_2	X_3
CPUC	P_4	X_1		
State legislature**	P_5	X_1	X_2	

Key decisions:	(1) Initial legislation introduced which included B^1, B^2, and B^5. (2) Final legislation passed which incorporated B^1, B^4, and B^5.
Conclusion:	Passage of LNG Siting Act of 1977 (S.B.1081) which defined a custom-tailored siting procedure for LNG. Some features: — CCC nominates and ranks sites in addition to the one applied for — CPUC selects a site from the CCC-ranked set, not necessarily the top-ranked site.

*CEC = California Energy Commission
**Interested party with responsibility for decision(s).

This behavior relates to the process described by Braybrooke (1978), where he points out that the issues are frequently transformed over time. Round B is a good illustration of this process. The problem was redefined into two new questions: How should need for LNG be determined? If need is established, how should an LNG facility be sited? The round was thus initiated when pressure to change the siting procedure was brought to the state legislature by the utility companies, the business community and the labor unions in California. Table 4 depicts the relevant alternatives which formed the basis for the debate on the elements of proposed legislation.

The industry and business interests saw the inevitable problem of obtaining local approval for a project in the national interest, but with costs to the local community. So the utility companies battled for a bill (S.B.1081) which would vest the CPUC with one-stop licensing authority, precluding any interference from local communities. The environmental and local interests, on the other hand, objected to a one-stop licensing process and favored a bill which required remote siting.

The resulting legislation was a compromise between the environmentalists, who supported consideration of offshore sites, and those who saw an urgent need for an LNG facility to assure energy and jobs. The CPUC was chosen over the more conservation minded CCC or the California Energy Commission as the agency with state permit authority, pre-empting local governments. As a bow to the conservationists, the CCC was given the mandate to choose and to rank possible sites, and to pass these rankings on to the CPUC. It was agreed that the site would not be offshore, as some environmentalists wished, nor could it be in a populated area, as the gas utilities wished. Indeed, an unpopulated area was strictly defined. There could be no more than an average of 10 people per square mile within one mile of the terminal, and no more than 60 people per square mile within four miles of the terminal.

The passage of the Siting Act of 1977 (S.B.1081) opened up a new procedure for finding an acceptable site and led to round C with the following problem formulation: Which site should be approved? The round was initiated by the CCC which, after considering 82 sites meeting the remote siting constraint, ranked the top four sites, Camp Pendleton, Rattlesnake Canyon, Point Conception, and Deer Canyon, in that order, on the basis of seismic, soil, wind and wave conditions, rough cost, and coastal resource considerations.*

These four alternatives form the background for the interaction among the interested parties in round C, as shown in Table 5. The CCC passed these rankings on to the CPUC which chose, by process of elimination, Point Conception, on the grounds that the two higher-ranked sites would involve unacceptable delay and would cause unacceptable risk to transients (i.e., campers, swimmers, etc.) at the nearby beaches and public parks. The CPUC, however, could only conditionally approve Point Conception subject to the utility company's ability to show that earthquake faults discovered in the area presented an acceptable risk to the terminal.

At the federal level, the FERC staff determined that the risks of both Oxnard and Point Conception were acceptably low, so that Oxnard should be preferred on land-use grounds; however, the FERC, choosing to avoid a federal-state confrontation, ruled in favor of Point Conception. After an appeal by the environmental and local interests, the Washington, DC Court of Appeals remanded the case back to the FERC on the ground that not all available seismic risk data were considered by the FERC in its ruling. This decision concluded round C.

Round D is still in progress at this time. As shown in Table 6, the initiating proposal is determined by the activities in round C which frame the alternatives as simply whether or not to declare the Point Conception site seismically safe. Only two parties, the FERC and the CPUC are currently active in the process, and they are considering only one attribute—the seismic risk at Point Conception. A final decision will depend upon whether the new studies show this risk to be above or below some acceptable level.

---

*Point Conception was included in the candidate set because S.B.1081 required that the applied-for site be included.

Table 5. Elements of round C.

Problem formulation:	Which site should be approved?					
Initiating event:	Applicant files for approval of Point Conception (the only site of the original three meeting the remote siting constraint of S.B.1081).					
Alternatives:	(Sites nominated by CCC plus applied-for site)					
	Camp Pendleton:	C^1				
	Rattlesnake Canyon:	C^2				
	Point Conception:	C^3				
	Deer Canyon:	C^4				
Interaction:	Involved parties		Attributes used for arguments			
	Applicant	P_1	X_1			
	FERC*	P_2	X_1			X_5
	CCC	P_3			X_3	
	CPUC*	P_4	X_1	X_2		
	Sierra Club	P_7			X_3	X_5
	Local citizens	P_8				X_5
Key decisions:	(5) The CCC has the following preference: $C^1 > C^2 > C^3 > C^4$ (6) The CPUC approval conditional on whether or not the seismic risk is acceptable. (7) The FERC consider C^3 acceptable. (8) Court requires FERC to consider additional data to determine whether or not seismic risk at C^3 is acceptable.					
Conclusion:	FERC and CPUC to consider additional seismic data.					

*Interested party with responsibility for decision(s).

Table 6. Elements of round D.

Problem formulation:	Is Point Conception seismically safe?		
Initiating event:	FERC and CPUC set up procedures to consider additional seismic risk data		
Alternatives:	Declare Point Conception safe:	D^1	
	Declare Point Conception not safe:	D^2	
Interaction:	Currently active parties		Attribute considered
	FERC	P_2	X_3
	CPUC*	P_4	X_3
Key decisions:	None yet. Future hearings are to determine whether or not seismic risk is acceptable for Point Conception.		

*Interested party with responsibility for decision(s).

4. INTERPRETATION OF THE MAMP MODEL: THE ROLE OF RISK ANALYSIS

A great deal of attention has been paid recently to the topic of technological risk assessment for problems such as the siting of facilities (see Conrad 1980, Schwing and Albers 1980). It is of interest to examine, in the context of the MAMP framework, the role that risk assessments have played in the California LNG case.

The sequential nature of the decision procedures, as clearly demonstrated by the increasing concreteness of the problem formulations through the four rounds of discussions in California, limits the possibilities for comprehensive analyses. The risk studies were carried out, not as an input to a broad energy analysis in California, but to support a more narrowly defined problem (Should site *x* or site *y* be approved?). Since round A in California was *not* defined in these narrow terms (the question of whether the terminal was needed was yet to be resolved), the analyses were ill-suited to address fully the issues on the table. In some sense, then, analyses designed to address the question of safety were prematurely introduced into a process that had not resolved higher-order questions of energy policy. Though they served to focus the debate on the safety question, they could not offer (nor were they intended to offer) a panacea for the resolution of the siting question.*

During the course of the LNG debate in California, six studies assessing the safety risks of the proposed terminals were conducted by the utility and local, state, and federal government agencies (for a critical review of these studies, see Mandl and Lathrop 1981). Several studies are of particular interest. The applicant commissioned a consulting firm, Science Applications Inc. (SAI), to do a study and the FERC produced its own risk assessment. Both reports showed very low numbers on various probabilistic measures of risk (expected fatalities per year and individual probability of fatality per year). These numbers were interpreted to mean that the risk was acceptable. A risk assessment produced by the consulting firm Socio-Economic Systems (SES) for the Oxnard municipal government suggested similarly low probabilistic measures of risk (though expected fatalities were 380 times higher than the applicant's assessment), but they interpreted the figures as unacceptably high.

One explanation for these different interpretations lies in the format for presenting the results. The SAI study described maximum credible accidents (MCAs) without accompanying probabilities. Opposition groups interpreted these results as evidence that the terminal was *not* acceptably safe. The municipal government originally in favor of the site, began to waver in its support, probably influenced by the apparent uncertainty of the risk and the strength of the opposition groups (Ahern 1980). In sum, the risk assessments did not provide a single, coherent assessment of acceptability of the risk of an LNG terminal; their results were subject to interpretation depending on party positions (Lathrop 1980). In fact, the risk assessments were used both to promote and to oppose terminal applications.

In reviewing the technical differences between the assessments leading to these conclusions, Lathrop and Linnerooth (1982) have shown that there are many degrees of freedom left to engineering and analytic judgment, including how to characterize risk, what formats to use for presentation, what gaps to fill with assumptions, which of several conflicting models to use, how to portray the degree of confidence in the results, and what contingencies to leave out of the analysis.

This analytic freedom helps explain the differences between the above three Oxnard risk assessments. It can push the risk measurement in any direction. Very conservative assumptions can drive it up; omissions of inconvenient aspects such as terrorism can drive it down. Clear presentations of expert disagreements can decrease the confidence in the results; and so on. The final result may have as much to do with the predilections of the analyst as with the physical characteristics of the site or technology.

---

*It is not surprising, then, that round A ended in a stalemate. The second round, where the state legislature took center stage, narrowed the problem (by resolving the question whether California needed a site) to one more receptive to technical risk studies.

This finding takes on special significance when viewed in the context of the policy process. The MAMP model has illustrated that the risk assessments, though intended to advise a client on the safety of the proposed terminal, were typically used to support a party argument. For this reason, clear incentives exist for the analysts to present their results as persuasively as possible, which explains the tendency on their part to omit discussions on the uncertainty of their results and to choose presentation formats that present their case as strongly as possible.

5. SUGGESTIONS FOR FUTURE RESEARCH

In this paper, I have attempted to give a flavor of the work of the Risk Group by describing the decision process in the USA in the context of the MAMP framework. The MAMP framework should be viewed as a starting point for undertaking research which can improve the political process with respect to problems such as the siting of facilities. We have seen that formal risk analyses, especially risk assessments, are subjective exercises undertaken to support a specific party's arguments. Furthermore, the importance of these analyses will depend on the nature of the sequential decision process, the relevant interested parties which interact, and the type of conflicts emerging.

Given these descriptive observations, several research areas appear to be promising avenues for the future. In a recent paper Nelkin and Polak (1979) indicate the inadequacy of existing institutions to deal with problems of conflicting evidence and polarized expert opinion with respect to questions such as risk assessments. As a way of dealing with this problem, they advocate the need to establish rules of evidence as a basis for making better decisions. Lathrop and Linnerooth (1982) provide a suggested set of guidelines with respect to establishing rules of evidence. In particular, they stress the importance of defining the risk being assessed, being clear on assumptions and error bounds as well as indicating the conditional nature of specific analyses which are undertaken.

There is a need for more field research which attempts to apply these criteria or others to a specific set of problems. One of the difficulties which currently exists is the lack of an institutional mechanism for evaluating the different risk assessments produced by different parties. Ackerman *et al.* (1974) point out that the traditional approaches such as legal responses, agency hearings and judicial reviews have inherent limitations with respect to evaluating these conflicting assessments. The problem is especially difficult for the siting of new technologies where there are no objective measures of risk. Private consulting firms frequently undertake these analyses but have a built-in bias in telling the contracting party what they want to hear.

With respect to the more direct consequences of siting a new facility O'Hare (1977) has proposed a compensation system to deal with opposition to proposed sites from certain interested parties. For example, suppose residents of a community are concerned with suffering losses in property values as well as safety and environmental risks if the project is sited near them. O'Hare proposes that each community determines a minimum level of per capita compensation for it to be willing to make a legal commitment to having the project in its backyard if the compensation is paid.

From the above suggested topics it should be clear that there is considerable research on risk which needs to be undertaken of a prescriptive nature. The purpose of our cross-country comparisons of LEG siting decisions is to provide data on how the political process appears to work in practice and the differences across countries. The MAMP framework described in this paper has been found to be a useful framework for making comparisons among countries. The challenge for the future is to capitalize on our understanding of process to try and improve political decision making.

REFERENCES

Ackerman, B., Rose-Ackerman, S., Sawyer, J., and Henderson D. (1974) The Uncertain Search for Environmental Quality. New York: The Free Press.

Ahern, W. (1980) California meets the LNG terminal. Coastal Zone Management Journal 7:185-221.

Allison, G.T. (1971) Essence of Decision. Boston: Little, Brown and Company.

Braybrooke, D. (1974) Traffic Congestion Goes Through the Issue Machine. London: Routledge and Kegan Paul.

Braybrooke, D. (1976) Policy formation with issue processing and transformation of issues, in: Hooker, Leach and McClennen (Eds.) Foundations and Applications of Decision Theory. Dordrecht, Holland: Reidel.

Conrad, J. (Ed.)(1980) Society, Technology and Risk Assessment. London: Academic Press.

Kunreuther, H., Lathrop, J., and Linnerooth, J. (1982) A Descriptive Model of Choice Siting Facilities. WP-82-18. Laxenburg, Austria: International Institute for Applied Systems Analysis.

Lathrop, J. (1980) The Role of Risk Assessment in Facility Siting: An Example from California. WP-80-150. Laxenburg, Austria: International Institute for Applied Systems Analysis.

Lathrop, J., and Linnerooth, J. (1982) The role of risk assessment in a political decision process, in: P. Humphreys and A. Vari (Eds.) Analysing and Aiding Decision Processes. Amsterdam: North-Holland.

Linnerooth, J. (1980) A Short History of the California LNG Terminal. WP-80-155. Laxenburg, Austria: International Institute for Applied Systems Analysis.

Majone, G. (1979) Process and outcome in regulatory decisions. American Behavioral Scientist 22:561-83.

Mandl, C., and Lathrop, J. (1981) Assessment and Comparison of Liquefied Energy Gas Terminal Risk. WP-81-98. Laxenburg, Austria: International Institute for Applied Systems Analysis.

Nelkin, D., and Pollak, M. (1979) Public participation in technological decisions: reality or grand illusion. Technology Review, September pp55-64.

Office of Technology Assessment (OTA)(1977) Transportation of Liquefied Natural Gas. Washington, DC: Office of Technology Assessment.

O'Hare, M. (1977) Not on my block you don't: facility siting and the strategic importance of compensation. Public Policy 25:409-58.

Schwing, R.C., and Albers, W.A. (Eds.)(1980) Societal Risk Assessment: How Safe is Safe Enough? New York: Plenum.

Science and Policy: Their Separation and Integration in Risk Analysis

Howard Raiffa

Graduate School of Business Administration, Harvard University, Boston, USA, and Chairman, Committee on Risk and Decision Making (CORADM)

The Committee on Risk and Decision Making (CORADM) was commissioned by the Division of Technology Assessment and Risk Analysis (TARA) of the US National Science Foundation to give advice on their research program in risk analysis. CORADM is an *ad hoc* committee in the Assembly of Behavioral and Social Sciences (ABASS) of NAS/NRC, and our report will be submitted to ABASS and TARA at the end of 1980.

The Committee members felt that if we are to report on research priorities for TARA we should first step back and reflect on how our society generates and copes with risk. Thus we have decided to prepare an extensive prologue to the report on research suggestions for TARA, and I should now like to discuss the stage reached in our deliberations on that prologue.

CORADM is primarily concerned with the adverse effects of risks to health, safety, and the environment. The following list illustrates a set of problems, which is far from exhaustive, that motivates our concerns:

- Three Mile Island
- Chemical waste
- Oral contraceptives
- Occupational exposure to benzene, asbestos, vinyl chloride, cotton dust, coke-oven emissions
- Alcohol consumption and cigarette smoking
- Accidental or intentional nuclear war
- CO_2 and atmospheric warming
- Extinction of plant and animal species
- Depletion of the ozone layer
- Health effects of dietary factors
- Acid rain
- New passive restraint standard
- Fluoridation
- DC-10s
- Air pollution (sulfates and small particulates)
- Decline of new prescription drugs
- Natural hazards (earthquakes, floods, etc.)
- Low-level radiation
- Nuclear high-level wastes

I will now outline what our committee *currently*—and I stress currently because we are still in a state of flux—is thinking about in terms of a prologue or a general report. After describing the structure and purpose of CORADM, the report will outline the macro-perspectives of risk by looking at the rising tide of public concern and document this by means of opinion polls, legislative and agency activity, public and private sector investments, media publicity, and increased litigation. Using time series of different indices such as infant mortality, premature death (i.e., death before 65), longevity, accidents, disabilities, sick days, birth

---

*This paper is an edited transcript of the talk given at the Sixth Symposium on Statistics and the Environment, October 1980.

defects, etc., the report will attempt to take stock of how our society is faring at the present time. These studies will be concerned not only with aggregate values, but also with the smaller-scale differences between various segments of our society. For example, there are still inequalities between blacks and whites, and between the US and other countries, and these represent potentials for improvement.

We will try to point out some of our greatest worries and potential problem spots, but, as is to be expected, committee members have some disagreements on this topic. Next we try to rationalize why public concern about risk has increased.

In the second main section of the report we discuss processes for managing and coping with risks. The word management often evokes different responses, so perhaps the term "risk management" should be defined more clearly. "Risk" is used differently in various disciplines such as in finance, economics, insurance, decision theory, engineering, etc. Risk management in its broadest sense involves the identification, estimation, assessment, monitoring, evaluation, and control of risk, including preventive, reactive *ad hoc*, and unorganized processes to deal with them. Institutions for this purpose include courts, legislatures, administrative agencies, business enterprises, labor unions, research institutions, citizen groups, and educational institutions, as well as individuals. Risk analysis itself can be subdivided into the assessment of uncertainties; and evaluation, i.e., policy analysis (other than assessment) for risk management.

Before discussing ways of coping with risks we look at ways in which they are generated in our society; i.e., self-imposed risks, co-generated risks, risks arising from productive activities or business and government, natural hazards, and risks that arise from the very institutional framework of our society. In this section we also discuss the slippery distinction between voluntary and involuntary risks.

Policy instruments for coping with risks can include, for example, information, incentives/disincentives, prohibitions, liability, transfer rights to generate risks, mediation and arbitration, insurance policies, regulation, etc. Most of these are self-explanatory, except perhaps for mediation and arbitration: for example, it is possible that in certain circumstances labor unions and businesses could jointly agree on means to control risks and then to obtain a formal stamp of approval on these contracts by government.

We next look at the various federal agencies that have to cope with risk and the different statutes that govern them. Statutes can be classified according to those that look at risks only (e.g., the Delaney Clause); those that try to reduce risks up to technological feasibility; and those that attempt to balance adverse consequences with benefits broadly interpreted. In practice, these statutes are not administered literally so that the distinctions are even more blurred than they first appear.

We look at what is happening in the courts, their levels of activity and different philosophical approaches, and how some commentators view their role in risk management. In addition to a discussion of public perceptions of risk and how they are formed, we want to discuss the problematic issues such as (i) what roles do (and should) perceptions play, and (ii) the thin line between education and indoctrination.

The most extensive part of our report concerns the roles of risk analysis, which may be summarized as follows:

Should keep in mind complex, socio-economic-political interactive processes for coping with risk
Most decisions based on common sense "ordinary" knowledge, with little formal analysis
Analysis can help with incremental choices
Analyses are often multi-purpose for multiple audiences
Analysis cannot eliminate judgments about uncertainties and values
Analysis can raise the level of discourse
Analysis can generate creative alternatives
Poor analysis may be worse than none
Analysis should be iterative: it should improve over time

Analysis can be used and misused as a political weapon
Analysis can be used and misused in the adversarial process
Need for peer review
Need for more and better analysts

Analysis can be subdivided into two categories, which can be called "in the large", and "in the small". An example of the former could be the case of an in-depth analysis done for an administrator of an agency who must decide (no action is a decision!) on what to do about a potentially troublesome chemical. It does not have to be an administrator or a chemical; the idea is that there is a high-level, (relatively) unitary decision maker who has a reasonably well defined problem. Possible solution settings are:

(a) Generation of policy alternatives (a checklist of things to think about)
(b) Consequences (impacts, costs, benefits)
  Health effects: how many? how much? who are they? how voluntary? how identifiable? etc.
  Non-health effects: to nature; to economic and business activity; to sociopolitical activity; to international relations
  Feedbacks of non-health effects on health effects
(c) Uncertainty analysis: assessment of uncertainties; natural science, behavioral science
  Being precise about degree of imprecision
  Volatility
  Disagreements
(d) Analysis for the dynamic decision process

Factors involved in the generation of policy alternatives include

Analysis can help generate creative alternatives
Sequential choices with intervening information
DM can sometimes collect information (e.g., experimentation)
Interdependence of problems (precedents)
Irreversibilities (physical, political, managerial—strict and partial)
Decisions about locus of action to be taken (e.g., level of government, decentralization through the market, etc.)
Institutional decision network (who has to decide what and when? Who has to be convinced, pressured, influenced, etc.)

The consequences of risk analysis can be summarized as follows:

Health effects:

1. How many people are (will be) affected?
   (a) in the entire population
   (b) in sensitive groups
2. How much are they affected?
   (a) mortality
   (b) morbidity
   (c) severe pain and suffering
   (d) psychological discomfort
   (e) anxiety
3. Who are they?
   (a) age distribution
   (b) income distribution
   (c) race/ethnic group
   (d) sex
   (e) occupation
   (f) geographical location
   (g) quality of life/health status

4. When will they be affected?
   (a) now
   (b) with some time lag
   (c) future generations
5. How voluntary/involuntary is the risk?
6. How "catastrophic" is the risk? (clustering of fatalities over time and space)
7. How "identifiable" are the victims (*ex ante* and *ex post*) and how "accountable" will the decision maker be?

Non-health effects:

1. Aesthetics
2. Effects on nature
3. Economic costs (and to whom)
4. Effects on economic growth, productivity, and innovation
5. Effects on business competition
6. Effects on other countries
7. Effects on distribution of incomes
8. Effects on public satisfaction with government

Secondary, tertiary, and general equilibrium considerations; net health effects: feedback fron non-health effects on health (e.g., 3, 4, 5, 7).

Next we turn to uncertainty analysis, and this will be given more detailed attention below. There next follows a category of activities which can be listed under the heading "the dynamic decision process".

Analysis for choice
Peer review of analysis
Adversarial inputs
Commitment to first steps
Public announcements
Adversarial documents
} Legal analysis

Influencing and persuading others
Educating others
Monitoring
Evaluating
Experimenting
Accumulating information
Guiding
Re-analyzing
Commitments to second steps
⋮

Another factor in the assessment of uncertainty, which we call "risk assessment" can be singled out for further discussion, as follows:

Why decompose?
Linkages between assessment and evaluation
Effective reporting of uncertainties (false precision or imprecision)
Special cases of assessments:
  rare catastrophic events
  carcinogenicity
  large systems (e.g., energy policy)
Need for judgmental synthesis in assessment of uncertainties:
  separation of facts and judgments about uncertainties
  separation of judgments about uncertainties and values
Criteria for effective reporting

Prudence in reporting *versus* prudence in action
Assessments by groups:
  spectrum of opinion
  why experts disagree; structuring
  vulnerability to external attack
Need for independent, credible assessment groups
Protection of scientific institutions; right to refuse to report
What to do if decomposition is uncomfortable

Figure 1 illustrates the decomposition of risk assessment and evaluation. Our report will treat at length the difficulties involved in this decomposition of tasks, and its advantages; if tasks are to be decomposed they still have to be linked and there is usually a lot of dysfunctional slippage in the linkages.

The uncertainties to be assessed can be roughly divided into two categories: those that fall within the natural science domain, and those in the behavioral or managerial science domain. Let us now turn to natural science uncertainties. Is a given chemical carcinogenic, for example? If so, what is its severity? What about exposures? What about health effects?

We next consider the effective reporting of uncertainties and the use of quantitative probabilistic reporting. We worry a lot about the false precision that comes with the use of numbers, and what to do about it, as well as the false sense of imprecision that comes with the use of qualitative reporting, such as with the use of such semi-quantitative terms as rarely, not unthinkable, beyond a shadow of doubt, not so often, sometimes, the preponderance of evidence shows, with some exceptions, etc. However, there are some advantages of probabilistic reports, as follows:

Less likely to be misinterpreted
Probabilities can be combined with other probabilities
Probabilities can be combined with losses and utilities—meshes with other factors
Probabilities can be compared (comparative risks)
Probability numbers are precise enough to be attacked

After discussing some special assessment problems we arrive at the conclusion that assessment is complex; only rarely do scientific facts speak for themselves. These disparate facts, theories, empirical findings, etc., have to be synthesized and made comprehensible for use in the policy making process (see Figure 2). This synthesis is required because:

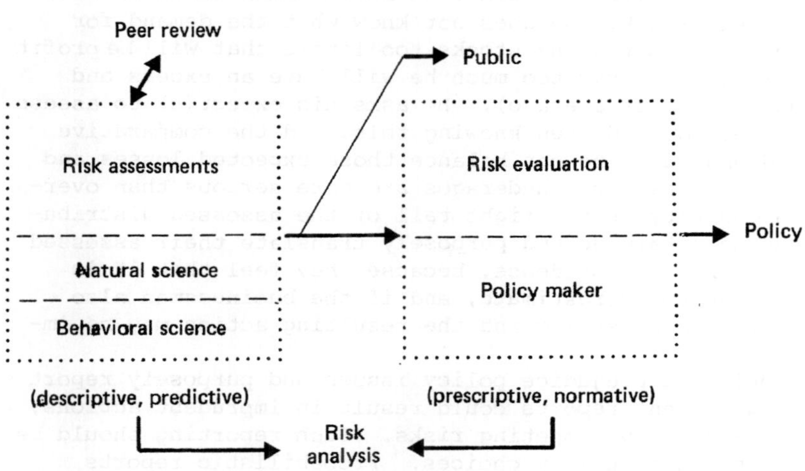

Figure 1. The decomposition of risk assessment and risk evaluation.

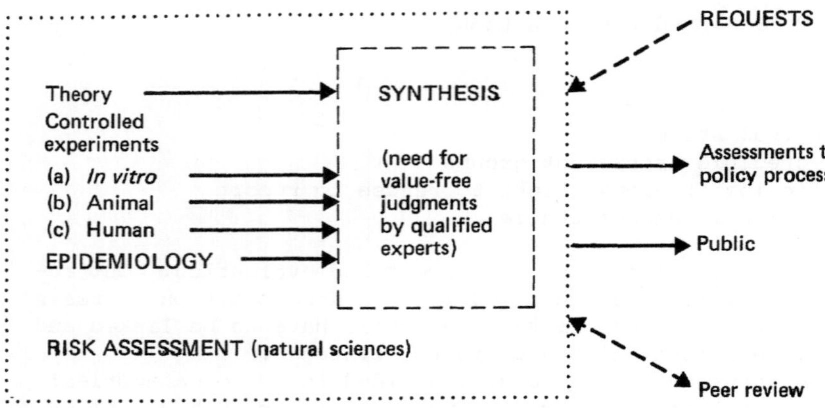

Figure 2. Synthesis of information for use in the policy making process.

Information comes from several sources, often conflicting, often messy;
Information is often very indirectly relevant;
Recitation of the set of indisputable facts may not be enough; may be useless to a nonexpert policy maker;
Judgments by experts may be indispensable for synthesis; but separation of fact and judgment, including value-free judgment, should be clearly demarcated.

The criteria for good reporting can be summarized as follows:

Inclusive—should synthesize relevant information (theory, experimental data, epidemiological data, etc.)
Free from the values of policy values—report should not prejudge policy conclusions; values appropriate to policy evaluation should not influence assessments of uncertainties
Comprehensive and meaningful to clients—informative and relevant
Useful in the decision process—use of proxy variables
Honesty and prudence

I would now like to dwell on the last criterion: honesty. This sounds nice, but there is a tension between honesty and prudence that should be clarified (see Figure 3), using an example from a business setting. A businessman wants to know how much of a given item he should stock. He does not know what the demand for that item will be, but he does know that if he stocks too little that will be profit foregone (a loss of underage); if he stocks too much he will have an excess and there will be profit foregone (a loss of overage). He asks his expert(s) to assess a probability distribution of demand, and then knowing this, and the comparative per unit losses of overage and underage, he can balance those expected losses and arrive at a decision. In the case where the underages are more serious than overages it is prudent to stock a quantity in the right tail of the assessed distribution. If the assessors of the uncertain demand purposely translate their assessed distribution to the right in the name of prudence, because they feel that it is more serious to underestimate than to overestimate, and if the businessman also compensates, then they may be overcompensating and the resulting action may be imprudent.

Probabilistic reports should not prejudice policy issues and purposely report with a prudent bias. Cascading prudent reports could result in imprudent actions, and there is a danger of double-counting competing risks. Such reporting should be honest, and not attempt to second-guess policy choices. Probabilistic reports about diverse consequences to health, for example, are very often slanted to be

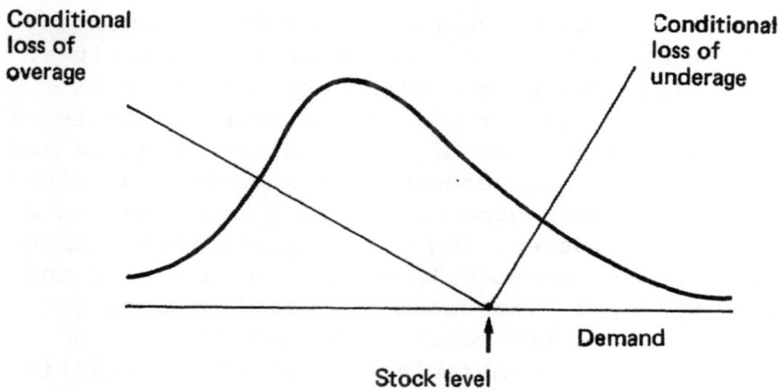

Figure 3. Prudence in reporting: an illustrative example. Optimal stock level is at a high fractile; the probabilistic assessment should not be distorted.

conservative. I believe that it is better to report honestly, and that prudence should, more appropriately, be accounted for in the evaluation process, rather than in the assessment process.

We will now look at assessments by groups. It is no secret that experts disagree; there are reputable scientists who hold fringe opinions at both ends of the spectrum—sometimes they are just plain wrong, and occasionally they are right. It is important that these fringe opinions are heard, but they should not be overweighted just because they are different. The media seek to inform and to interest their readers, so there is always a tendency for them to over-represent the tails of the distributions. Theory states that experts should agree, but they often do not, for the following reasons:

- Slippage in vocabulary
- Different experiences and difficulties with articulation
- Imprecise overlap of information
- Different paradigms
- Cognitive biases (anchoring, etc.)
- Elicitation biases
- Effect of role
- Personal interactions and reinforcements
- Conscious biases:
  - to help mankind — to compensate for adversarial purposes
  - to help oneself — money, not to curry disfavor of peers, colleagues, friends, children

When experts disagree it is important not to suppress their differences; and these differences, if they are to be understood and properly reflected in the policy process, should be structured. Do the experts agree on what they disagree about? Do they agree on what further analyses (modeling, data collection, etc.) would be desirable to bring their views into closer harmony? Attention should be given to meaningful ways to report disagreements; an eye should be kept on ways users of reports react, and should react, to such reports.

Many assessments (such as of adverse effects to health and the environment) are clear-cut and no elaborate quantitative analyses are required because the facts speak for themselves. When we concentrate discussions on those assessments that present difficulties, we should not lose sight of the fact that many are routine and that a little analysis without controversial judgments may go a long way. Some given chemical may be clearly carcinogenic, substitute chemicals may exist, and the given chemical may not yet be on the market. That's easy. Another easy case is where the chemical is given a clean bill of health.

While the bulk of cases studied may be routine, there are plenty of important ambiguous cases where the facts (about uncertainties) do not speak for themselves. Facts, theories, ambiguous experimental findings, uncontrolled epidemiological findings, etc., all must be synthesized by experts for policy analysis and decision by nonexperts. Experts, when they look at a given set of facts, might disagree and might organize their deliberations about such disagreements, but suppose that after considerable internal discussion a risk assessment group comes to a consensus on a synthesis of facts and judgments about uncertainties, and then reports this consensus to a risk evaluation group. This risk assessment will also be made public and reviewed by other experts, some of whom will not like the policy implications of the report. Those external reviewers may attack the conclusions and try to find vulnerable spots in the report; and what is more vulnerable than an only partially supported judgment? It may be the case that some members of the risk assessment group have consciously (and therefore inappropriately) biased their inputs, while others might have fallen prey to subtle subconscious biases. Such weaknesses should rightfully be adversely criticized.

External reviewers with strong *policy* viewpoints will often attack risk assessments when the implications are not congenial to their preferred policies. To be effective, those attacks should be concentrated on potential weaknesses of the synthesis process. Even if a member of the risk assessment group has meticulously tried to subdue his or her potential subconscious biases, this might not actually be perceived as being the case by an outsider; and even if it is perceived to be the case, the adversarial outsider might win debating points by casting suspicion on the integrity of the insider.

It is not easy for a risk assessment group, in a highly charged controversial domain, to fulfil its mission responsibly even if its assessments are to be privileged documents only for the eyes of the policy maker and his discreet staff. But when these assessments have to withstand the barbs of those with strong policy convictions the task becomes doubly difficult. It takes courage and an impeccably neutral committee member to say what he or she actually believes in a controversial case because that member henceforth will be branded as an exponent of a cause. There is a tendency for individuals, who jealously want to be perceived as neutral, to soften their true opinions and to bend over backwards to be fair to the other side. Thus another bias enters the scene.

There are lots of idealistic, dedicated scientists who are motivated to seek the "scientific truth", wherever that may lead them, and who are more than willing to engage in academic interchanges to seek out collective wisdom. They may realize that they are prey to subconscious behavioral biases and they might be consciously eager to monitor their actions to avoid egregious conflicts of interest. It is important to encourage scientific groups to foster a tradition of openness and honesty in reporting. Such groups would, if created, deserve acclaim, and financial support should be properly laundered in ways to bring a minimum of external pressure. I am not recommending anything as grandiose as the scientific court of respected elders, but hope that many panels and committees can be founded in universities, in consulting firms, in industry, in scientific academies, and in government agencies that take pride in their scientific integrity and work hard to fulfil their responsibilities. Something is wrong when the membership of prestigious committees, created to report on some natural science uncertainties and not on policy issues, must be structured to yield a balanced portfolio of policy viewpoints rather than a balanced portfolio of scientific and methodological skills.

The scientific community must understand the dilemma we are in. It is imperative both for the progress of science and for better decision making that scientific reports be openly available and subject to peer review. But this puts a burden on assessment groups that must mold together science and judgment. Science cannot simply stand aloof from all the policy needs of our society, but when it gets involved it should expect to be attacked and should carefully prepare its defenses. There is some safety when a multiplicity of scientific institutions engage in these controversial activities and more groups within existing institutions should be created and suitably financed to undertake such studies. Occasionally it may be desirable for some such group simply to say that it is not ready to give its opinions on some controversial topics.

Even the Supreme Court prefers not to consider some legal issues because they are currently too divisive, just as the members of a risk assessment group may do. The reluctance of the Supreme Court to delve into some issues is justified by its grander role in society. Why should it jeopardize its mission by getting trapped into a morass of emotional conflicts when society is not ready for reasoned arguments. A similar argument can be made on behalf of science. When the scientific facts do not speak for themselves, when judgments are suspect, or when advocates are ready to pounce on all sides of an issue, then occasionally scientists and methodological experts should say that they are simply not ready to make pronouncements in this field under the auspices of such and such an institution. Scientists are individuals, of course, have the right, and some would say the obligation, to speak out. However, here I am making a distinction between scientists as individuals, and scientists who speak on behalf of scientific institutions.

This leads to a final point: what should be done if the decomposition of risk assessment from risk evaluation cannot be comfortably achieved? Analysis has to be much less formal and presumably the principal decision maker will need to get integrative advice that mixes facts, judgments about uncertainties, and opinions about values.

We now come to risk analysis "in the small", which can be summarized as follows:

- Analysis does not have to be "grandiose"
- Analysis for individual (small) actors (within a complex system) can be simple
- Sometimes a little analysis can go a long way
- Analysis may be too difficult and not worth doing
- General-purpose analyses can help many users (one reason why assessments of uncertainties should be isolated)
- Analysts should understand the way society copes with risk and be prepared to help "in the small" as well as "in the large"
- Analysis may not only help to "solve" problems, but also to devise new ways of thinking about old problems, and to generate new ones

The important message here is that analysis, if it is tailor-made for specific individual purposes, need not be horrendously intricate. Often a little analysis can go a long way, because, for example, a given actor may be severely constrained in his choices; many attributes of concern that one can think of *a priori* may yield roughly equivalent outcomes; and only one or two uncertainties may be of prime importance.

Next we come to value controversies, and here, as elsewhere, our intention is to identify some key problems rather than to solve them. The following list is far from complete.

Value controversies

- Trade-offs of incommensurables:
  - Expenditures for life saving
  - Allocations with fixed budgets
  - Size of budget
  - Shadow prices of life saving by agencies
  - Should shadow prices be uniform?
  - American lives versus lives of others (foreign aid)
- Trade-offs within domain of health
  - Mortality/morbidity/psychological well being
  - Lives, ELYs, QUALYs, EDRAs
- Temporal trade-offs
  - Discounting (of lives?), intergenerational trade-offs; lives of "others" today versus lives of Americans in the far future
- Efficiency equity
  - Project-by-project equity, cyclical equity, income redistribution

Identifiability and accountability: probabilistic identifiability, *ex ante* and *ex post*

Protection of most sensitive groups with self-imposed sensitivities (e.g., air pollution and the heavy smoker, people living in the flood plains, etc.)

Clustering of deaths over space and time (e.g., 500 deaths at once, versus 600 isolated deaths)

Paternalism: arguments for and against

Libertarian views on imposed risks

"Man" versus "nature"

Rights of the fetus

⋮

We start off with the popular controversy: dollars or lives. It is interesting to estimate the marginal cost of life-saving activities in the various federal agencies. The values range from tens of thousands to tens of millions of dollars per life saved. Most commentators who are bothered by the idea of placing a dollar value on anyone's life will nevertheless feel it important to examine how a given budget would be efficiently spent on saving lives. I have parenthetically included in the above list a remark about saving foreign versus US lives because we have the policy option of doing something dramatic to save foreign lives (e.g., those starving in the Sahel).

But we also have problems of trade-offs within the domain of health; such as between mortality, morbidity, and states of psychological well being. Instead of counting only numbers of lives saved, we might want to look at expected life years (ELYs) or quality-adjusted life years (QUALYs) saved. Various mortality states might be commensurate in terms of equivalent days of restricted activity (EDRAs). How these trade-offs are, and should be, made involves controversial value judgments. The remaining value controversies can also not be ignored in important policy settings.

The final sections of part III of the CORADM report (Roles of analysis) have hardly been discussed by the committee. Section 6, entitled "Analysis in the adversarial process" includes one topic that has been repeatedly pointed out to us by environmentalists, namely, that resources for anlysis are unevenly distributed across adversarial groups. For instance, business is most highly endowed, and environmental and public interest groups have little resources. The situation may not be that unbalanced because a lot of academic researchers are antagonistic to business, and many are willing to work for a cause. Nevertheless, there is a perceived imbalance, and some environmentalists recommend that government agencies should support the analytical capabilities of groups that are challenging the "system".

For example, Wash 1400 (the Rasmussen Report) was a mammoth multi-million dollar study that was comforting to the nuclear industry. Scientific groups that did not like its findings attacked it and looked for flaws—and they found many. Should that study, to be credible, have been undertaken at the outset by two independent groups? Or should funds have been set aside for external peer reviews that would not have been designed to be adversarial? I have my own opinions on these questions, but our committee has not yet debated them. The trouble is that there is too much debate, and the more we probe, the more we find fine points on which to disagree.

The prologue of our report to TARA can be outlined as follows—as perceived today, but not necessarily as of tomorrow. All this comes before our research recommendations to TARA.

PROLOGUE TO THE REPORT TO TARA

Chapter 2

RISK ASSESSMENT IN A PROBLEM CONTEXT

Comparing Risk Assessments for Liquefied Energy Gas Terminals – Some Results

Christoph Mandl and John Lathrop
International Institute for Applied Systems Analysis, Laxenburg, Austria

1. INTRODUCTION

1.1. Purpose, Background and Scope

One of the most challenging problems in decisions concerning the deployment of novel, large-scale technologies is the assessment of the risk to the surrounding population. In particular cases, such as nuclear reactors or liquefied energy gas (LEG) facilities, the political process involved may tend to focus on one particular form of that risk, i.e., the risk to life from catastrophic accidents. This paper examines several different assessments of that type of risk with two main goals in mind:

(i) To present and compare the various procedures of risk assessment as applied to liquefied energy gas (LEG) terminal siting, and in doing so to clarify the limits of knowledge and understanding of LEG risks.
(ii) To quantify and compare the risks at four LEG terminal sites: Eemshaven (Netherlands), Mossmorran (UK), Point Conception (USA), and Wilhelmshaven (West Germany).

In the last decade a new technology for transporting and storing natural gas has become increasingly important for the overall energy supply of industrialized countries. The central idea of this new technology is to reduce the temperature of natural gas below -162°C, at which point natural gas becomes a liquid with one six-hundredth of the volume of the gas. The advantage of liquefied natural gas (LNG) is that it can be transported and stored efficiently in tanks due to its high energy per unit volume. Only in liquefied form can natural gas be transported via ships at reasonable cost.

However, due to the extremely low temperatures of LNG it is necessary to build special ships, special terminals to transfer LNG to and from the ships, and special tanks on land to store LNG. Cost considerations have made it necessary to plan and build LNG vessels and LNG terminals of considerable size; a typical vessel can contain 125 000m^3 LNG. At LNG terminals up to 60 000m^3 LNG is transferred per day and terminal storage tanks are planned to contain up to 500 000m^3 LNG. It is therefore not surprising that this high concentration of LNG at the site of a terminal has created concern that there might be potential negative effects, particularly to the environment and to the local population.

This paper in fact covers a broader category of terminals than those handling LNG. One of the terminals examined is to handle liquefied propane and butane. While LNG is stored at -163°C at very low pressure over ambient, liquefied propane and butane are stored at much higher temperatures and pressures, leading to significantly different behaviors during spills. However, all three substances involve essentially the same accident scenarios, though with different parameters and probabilities of detonation. Consequently, propane and butane have many of the same risk assessment features and problems as LNG. Since all of these substances, liquefied methane, propane, and butane, are called liquefied energy gases (LEG), the terminals examined in this study will be referred to as LEG terminals. Although there are many aspects involved in assessing the advantages and disadvantages of an LEG terminal at a specific site, the risk to the local population has turned out to be a crucial question. Because of the lack of historical data on accidents at LEG terminals, the frequency of such accidents as well as their consequences to people cannot be readily estimated. Therefore, over the past seven years attempts have been made to quantify the risk to the local population for different planned LEG terminals, using different techniques and models, with different results.

It is the purpose of this paper to review carefully the risk assessments undertaken for four LEG terminals in four countries, to discuss their plausibilities, explain their differences, compare their risk estimates and draw conclusions concerning their usefulness and limitations. Where necessary and appropriate we also expand some of the risk assessment reports. While this is not the first comparison of LEG risk assessments (see, e.g., HAZEL-REV; see references for explanation of acronyms), it is the first that we know of to compare assessments from different countries.

Because LEG terminal risk assessment is a new technique, there is still some disagreement among experts as to how to quantify risk, which models to use, what to include in a risk assessment, and what to exclude. Clearly, no pretense is made that this report provides complete or final answers concerning comparative risks or risk assessments. Rather, it describes some initial attempts to address important problems in the field of risk assessment.

1.2. Risk, Probabilities, and Consequences

Before it is possible to quantify risk, we must define it. It will become apparent in this section that different people mean different things when they talk about risk. Therefore our definition (actually a set of definitions) cannot be descriptive but rather will be prescriptive.

Ideally, if one adopts the axioms of rational choice under uncertainty, the evaluation of any decision alternative should consider the probability distribution over the consequences resulting from the alternative, which may be expressed in a space of several dimensions (see, e.g., LUCE-GEN). Yet the concept of risk singles out a subset of those dimensions for special analysis. The term is typically applied to particular uncertain costs, diverting attention from other costs and uncertain benefits that could be just as important for evaluation. In the case of LEG importation, for example, several dimensions are of concern for site selection and facility design. Of those, several involve uncertain costs, such as financial losses to the applicant if anything goes wrong (delay in application approval, loss of source contract, ship accident); environmental losses due to accidents or even routine disruption; fatalities and injuries due to accidents; property losses due to accidents; and losses to consumers from natural gas supply interruption if anything goes wrong (these could include unemployment and health effects in an extreme winter). While all of these uncertain costs could be and are referred to as risks, and all of them could be analyzed by techniques of risk assessment, in fact the term risk assessment in the context of LEG typically refers only to assessments of uncertain loss of life due to accidents. That is the scope for all risk assessments reviewed in this paper.

The best way to develop a definition of risk is to start by quoting some of the definitions from the risk assessment literature.

SAI-USA: "Risk is the expected number of fatalities per year resulting from the consequences of an accidental event".

CREM-UK: "Risk is the probability of an injurious or destructive event, generated by a hazard, over a specified period of time".

BATTE2-OTH: "Group risk is defined as 'the frequency at which certain numbers of acute fatalities are expected from a single accident'. The risk to society as a whole is defined as 'the expected total numbers of acute fatalities per year resulting from accidental events in the system'".

Surveying the set of risk assessments reviewed in this study, one can identify two extreme definitions of risk. One extreme, given by the risk definition of CREM-UK considers probabilities of destructive events only, and does not look at the consequences these events can have. Such an approach only makes sense for comparison or evaluation in the very limited case when all destructive events have equally valued consequences, and risk is defined as the probability that any one of the events will occur in a given time interval. It would be clearly meaningless to label two facilities equally risky if they had equal probabilities of an accident, but where an accident in one facility would have much more serious consequences than an accident in the other facility.

At the other extreme, risk can be, and sometimes is, viewed as the worst possible event (with the most serious consequences). Again we would argue that comparing this kind of risk is not meaningful because it omits the probability of an event and thus the relevance of such a worst possible event. We thus find that the definitions of risk described by Keeney *et al.* (KEEN-OTH) are the best prescriptive definitions:

(i) *risk of multiple fatalities:* probability of exceeding specific numbers of fatalities per year;
(ii) *societal risk:* total expected fatalities per year;
(iii) *group risk:* probability of an individual in a specific exposed group becoming a fatality per year;
(iv) *individual risk:* probability of an exposed individual becoming a fatality per year.

Each of these definitions addresses a different aspect of the political perspective of risk.

Risk of multiple fatalities is typically displayed as a complementary cumulative probability distribution: the probability per year that the number of fatalities will exceed x shown against x. Such a curve, sometimes called a Rasmussen curve, contains information not available in the individual probabilities: the effect of correlations between those probabilities. A Rasmussen curve addresses the sensitivity to catastrophe found in the political perspective of risk. Consider two facilities that cause equal numbers of expected fatalities per year: in one facility those are bunched into very rare catastrophes, and in the other they are spread over common small accidents. The former facility may encounter greater political opposition due to sensitivity to catastrophe.

Expected fatalities per year is appropriate for particular types of analysis, such as cost-benefit or risk-benefit analysis, where social preference is assumed to be linear in terms of number of lives lost.

The third definition (the probability of an individual in a specific group becoming a fatality per year) could be used to address the sensitivity toward equity found in a political perspective of risk. This measure enables one to determine in some sense how much of the risk is being borne by neighbors, campers, boaters, etc. This definition also allows separate determinations of occupational and non-occupational risks, two risks which are often treated quite differently in political and social processes.

Risk, according to the fourth definition (the probability of an exposed individual becoming a fatality per year), is simply an average over the group risks measured by the third definition. This measure is somewhat troublesome because it is dependent on the definition of exposed population. If "exposed" is defined as having an individual probability of fatality of greater than 10^{-12} per year, the individual risk will be averaged over a region extending not too far from the facility. If, on the other hand, "exposed" is defined with a cut-off probability of 10^{-30} per year, the individual risk will be averaged over a much larger region, and will be much lower. In spite of this shortcoming, individual risk is a measure that allows a convenient comparison of the measured risk and more routine risks the individual may face (e.g., the risk due to smoking, driving, etc.). While such comparisons do not fit into a decision or choice framework, they do provide readily understandable benchmarks for scaling the risk of a facility.

1.3. LEG Terminal Risk Assessment as a Decision Aid

Given the orientation of this paper, it is easy to forget that a risk assessment is not an end unto itself, but is in fact only one element of the complex process of LEG facility siting and design. More importantly, a risk assessment is supposed to be a decision aid for one or more of the decisions that must be made within that process. An understanding of where a risk assessment fits within an LEG siting and design process is essential to the understanding of the adequacy and worth of a risk assessment as a decision aid. One aspect of that process is of particular importance here: risk to life is only one dimension of concern for the decisions involved. Other dimensions include cost, land use, environmental quality, air quality benefits of natural gas, and dependence on foreign sources. There are also other important dimensions involving risk: supply interruption risk due to shortage, embargo, accident, earthquake, or bad weather preventing berthing.

Given the many decisions involving risk dimensions that must be made, it would seem that there are several roles for risk assessment in LEG facility siting and design. Yet the processes studied in our research have narrowed that role down to a single application: in one dimension, risk to life and limb; at one level, siting or design (depending on the country). There are several effects of this narrowing. To begin with, it diverts analytical effort and political attention away from those questions not addressed by risk assessment. For example, supply interruption risk could be a significant factor.

A second effect of the narrow role given to risk assessment is that the level at which it is applied affects how it is conducted. When risk assessment is part of the site selection process, a particular facility design is assumed, and analytical effort concentrates on such things as shipping traffic and local population density as site-specific inputs in a calculation of population risk. When risk assessment is part of the facility design process the site is assumed to be fixed, and the analysis considers the sizes, arrangements and specifications of components of the facility. In that case technical variations on the design are considered in terms of incremental reductions of risk.

There is a third effect of the narrow role assigned to risk assessment. Once a site is selected, given the political realities of the situation, the question of the overall acceptability of the risk is more or less settled. If a risk assessment is applied at the design level, it may consider various modifications to reduce the risk in the most cost-effective way. However, given its scope and charter, the assessment is highly unlikely to find that the site cannot be made acceptably safe with current technology and so should be abandoned. On the other hand, if a risk assessment is applied at the site selection level, it would at least be feasible to rule that none of the sites in the current choice set is acceptable.

Risk assessment does not exist in a vacuum. It is a decision aid in a much larger process. Any understanding of current assessment, and any suggestion for improvement, requires an understanding of that larger process. As this section has pointed out, that larger process controls the role and nature of risk assessment in very basic and important ways, even though the assessments may be carried out as strictly independent studies.

2. REVIEW OF RISK ASSESSMENT REPORTS

Table 1 gives a comprehensive overview of the most important risk assessment reports available to us, including not only those prepared for the four sites—Eemshaven, Mossmorran, Point Conception, and Wilhemshaven—but also a few others of particular interest. Some comments on the row headings of this table might be helpful.

(a) *Parts of the system considered.* Not all reports consider all the main parts of an LEG terminal, namely the vessel, transfer, and storage tanks. In particular, for Wilhelmshaven there are two types of report, one dealing only with vessel operation and LEG transfer, and the other dealing only with the storage tanks.
(b) *Concept of risk.* As discussed in Section 1.2, there is no unique definition of risk. Each report should therefore be quite specific on what type of risk is analyzed.
(c) *Estimation of probabilities of events.* One crucial part of risk assessment is the estimation of probabilities, unless only the consequences are considered. It is therefore necessary to see how this problem is solved in different reports. Two techniques can assist in performing this task for fixed plants. The *event tree* is a technique for developing a logical sequence of events (failures) resulting in unwanted consequences (accidents), and can help to avoid overlooking possible accidents. Having identified the possible events (failures), the goal of *fault tree analysis* is to identify and determine the probability of a "top-level event" (typically a specific accident) that would be the result of a sequence of basic events (failures) of the system. However, these techniques are not appropriate for estimating probabilities of accidents such as ship collisions. Two methods for estimating those probabilities are discussed later.
(d) *Estimation of consequences of events.* It is necessary for the consequences to be stated in terms a decision maker is concerned with. For this reason, and because of the definitions of risk typically assumed, most reports estimate the consequences in terms of the number of fatalities a certain event could cause.
(e) *Estimation of risk.* Different estimations are given depending on the definition of risk employed. In some cases no estimation is given at all.
(f) *Final findings.* The ideal result of a risk assessment report should be the quantification of the risk in comparison with risks from other sources such that the decision-making process can determine whether the risk from an LEG terminal is high or low compared with other risks. The ideal comparison is between risks from alternatives actually faced in the decision-making process: site A against site B, site A against no site, risk mitigation I against risk mitigation II, etc. Such a risk comparison is the risk assessment result of most direct usefulness to the decision process. In any case, it should be kept in mind that decisions concerning the acceptability of the risk from an LEG terminal involve social value tradeoffs and perhaps political considerations that go beyond the mission of the risk assessment and the legitimate authority of technical risk assessors. It follows that the final finding of a risk assessment should impart information to the decision maker for him to use as a basis for his decision without making that decision for him.
(g) *Uncertainties in final findings.* Due to the lack of experience with LEG accidents there remains a substantial amount of uncertainty about the accuracy of the estimations of probabilities and consequences of events. Different reports handle this problem differently: some ignore uncertainties completely, some give conservative estimations, some perform sensitivity analysis, and some give error bounds on the quantified risk.
(h) *Single event with highest risk.* If mitigating measures to reduce the risk are undertaken it is interesting to know which event bears the highest risk, as it is often the case that the highest-risk event offers the most cost-effective opportunities for mitigation.

When evaluating the reports one should keep in mind that the differences between the reports that become obvious from Table 1 can at least partially be explained by the fact that they were prepared and used for different decision processes and therefore each report was developed in a way suited to the particular decision process it was to serve.

Table 1. Comparison of reports on issues.

Issues		TNO1-NL	ACTION-UK	CREM-UK	ADL-USA	FERC-USA	SAI-USA
a	Parts of system considered	Vessel, transfer, storage tank	Vessel	Vessel, transfer, storage tank	Vessel, transfer, storage tank	Vessel	Vessel, transfer, storage tank
b	Concept of risk	Risk of multiple fatal-ities and group risk	Group and individual risk	Probability of an injurious or destructive event	Multiple fatalities risk	Societal, group, and individual risk	Risk of multiple fatal-ities, group & individual risk
c	Estimation of:						
c1	probabilities of events	Yes, quantitative	Yes, quantitative	Only in terms of low, very low, etc.	Yes, quantitative	Yes, quantitative	Yes, quantitative
c2	event tree analysis used	Yes	No	No	Yes	Yes	Yes
c3	fault tree analysis used	No	No	No	Yes	No	Yes
d	Estimation of consequences of events	Yes, quantitative in terms of fatalities	Yes, quantitative in terms of fatalities	Yes, but only physi-cal cons. (eg, spill size); no estimation of fatalities	Yes, quanti-tative in terms of fatalities	Yes, quanti-tative in terms of fatalities	Yes, quanti-tative in terms of fatalities
e	Estimation of risk	Societal & indi-vidual risk low cf. other man-made risks	Individual risk high cf. other man-made risks	No estimation of expressed fatalities; only of probabilities of events	Yes, quantitative	Yes, quantitative	Yes, quantitative
f	Final finding	Societal and individual risk low cf. other man-made risks	Individual risk high cf. other man-made risks	No reason to doubt that installations cannot be built and operated in such a manner as to be ac-ceptable in terms of community safety	Pt. Conception suitable with respect to vessel traffic safety. Risk is very low.	Risk compar-able to risks from natural events & thus on an accept-able level	"The risk is extremely low"
g	Uncertainties in final findings	Not mentioned	Not mentioned	Not mentioned	Sensitivity analysis	Disagreement among experts is mentioned	Sensitivity analysis
h	Single event with highest risk	Grounding of LNG tanks	Not identified	Not identified	Not identified	Not identified	Not identified

Table 1. Continued.

Issues	BROTZ-D	KRAPP1,2,3-D	WSD-D	BATTE-OTH	HSC-OTH	KEEN-OTH	SES-OTH
a	Vessel, transfer	Vessel	Vessel	Vessel, transfer, storage tank	Vessel, transfer, storage tank	Vessel, transfer, storage tank	Vessel, transfer, storage tank
b	Not defined	Not defined	Not defined	Multiple fatality, societal, and group risk	Multiple fatality and group risk	Multiple fatality, societal group and individual risk	Multiple fatality risk
c1	Only in terms of very low	Yes, quantitative	Only in terms of very low	Yes, quantitative	Yes, quantitative	Yes, quantitative	Yes, quantitative
c2	No	No	No	Yes	Yes	Yes	No
c3	No	No	No	No	No	No	No
d	Yes, but only physical cons. (eg, spill size); no estimation of fatalities	No estimation given	Some quantitative statements in terms of few and many fatalities	Yes, quantitative in terms of fatalities	Yes, quantitative in terms of fatalities	Yes, quantitative in terms of fatalities	Yes, quantitative in terms of fatalities
e	No estimation given	No estimation given	Yes, quantitative	Yes, quantitative	Yes, quantitative	Yes, quantitative	Yes, quantitative
f	With regard to consequences & their probability there is no danger, cf. relevant laws	No final findings	Risk is not insignificant	Risk about the same as that from the gas distribution network	Risk only acceptable if suggested mitigating measures are undertaken	Risk less than those that the population near terminal is exposed to presently	Level of safety cannot be specified accurately
g	Not mentioned	Not mentioned	Mentioned	Considered, & error bounds given	Not mentioned	Sensitivity analysis conducted to examine effects of variations of 2 parameters	Considered, & error bounds given
h	Not identified	Not identified	Not identified	Rupture of transfer pipeline with delayed ignition	Not identified	Not identified	Not identified

3. ASSESSMENT AND COMPARISON OF LEG TERMINAL RISK

In this section the probabilities and consequences of different events (failures) will be discussed. The procedure follows the line of the reports SAI-USA and BATTE2-OTH. After giving a technical description of the four LNG terminals the rest of this section is divided into three distinct parts. First, estimations of the probabilities of failures are considered, then the estimation of the size of a vapor cloud and its ignition probability as a result of the failures, and finally the consequences to the local population. The primary purpose of this section is to present the results from risk assessment reports in a comparable manner and to discuss important differences in estimates of probabilities and consequences between the reports in terms of the underlying assumptions of the models used and their plausibility.

However, as we have already shown in Section 2.1, not all reports are easily comparable. Some do not consider all the events discussed, while others do not quantify either the probabilities or the consequences of events. Therefore, this section cannot and will not be a complete comparison for all events.

In Table 2 we give a brief description of the planned terminals at Eemshaven, Mossmorran, Point Conception, and Wilhelmshaven. As can be seen, Mossmorran is a different type of terminal from the others. Not only is it an export terminal, but the exported gases are mainly propane and butane, while LNG consists mostly (approximately 90 percent) of methane. As far as one can tell from the available risk assessment reports, the technical layouts of the different terminals are much the same. Not only are the LEG vessels similar (except in size) or even the same, but also the storage tanks and the transfer systems are very much alike.

3.1. Events: Their Probabilities and Resulting Spill Sizes

One of the most difficult questions in risk assessment is the identification of possible events or failures and the estimation of their frequencies or probabilities. By definition it is almost impossible to obtain enough historical data to estimate the probabilities of a low-probability event. Rather, one has to build models and rely on data from other presumably similar systems. Another important part of the problem is the identification of events that have never occurred before that would have serious consequences. This problem was acknowledged in the Lewis Report (LEWIS-REV), where it was stated that:

> "It is conceptually impossible to be complete in a mathematical sense in the construction of event-trees and fault-trees; what matters is the approach to completeness and the ability to demonstrate with reasonable assurance that only small contributions are omitted. This inherent limitation means that any calculation using this methodology is always subject to revision and to doubt as to its completeness."

We therefore do not and cannot claim that the events considered here are a complete set of possible events. However, it can be said that this set of events includes all events that were thought of in the risk assessment literature, e.g., TN01-NL, SAI-USA, ADL2-USA, BATTE2-OTH. The two major failures of concern are vessel accidents and storage tank ruptures, both of which are discussed below.

Philipson (PHIL-GEN) describes two methods typically used to establish estimates of the probabilities of vessel accidents:

(i) *Statistical inference*. Estimates are computed using historical data, first for a larger class of ships, such as oil tankers, and then modifying the estimates to account for the anticipated differences in LEG ships and their operations at the specific harbor. This is done, for example, by employing judgment and by assessing the proportion of past accidents that would not have occurred if various capabilities of the system had been in place.

Table 2. Description of terminals and sites.

	Eemshaven	Mossmorran	Point Conception	Wilhelmshaven
Type of terminal	Import	Export	Import	Import
Type of transferred material	LNG	Propane/butane (liquefied) and gasoline	LNG	LNG consisting of 90% methane, 5% ethane, propane/butane
Average transfer per day (in m^3 liquefied or MW)	18 500 m^3 ≃4 900 MW	13 400 m^3	Initial: 58 500 m^3 ≃15 500 MW current plan: 41 000 m^3 ≃10 900 NW	56 500 m^3 15 000 MW
Maximum capacity of ships	125 000 m^3	60 000 m^3 propane/butane 10 000 m^3 gasoline	130 000 m^3	125 000 m^3
Number of ships per year	54	80 for propane/butane 9 for gasoline	190	170 ships of 125 000 m^3 264 ships of 10 000 m^3
Number and capacity of storage tanks	2x120 000 m^3	4x60 000 m^3 propane/butane 2x31 000 m^3 gasoline	2 later 3 with 77 500 m^3 each	6x80 000 m^3
Number of people living within 2 km of terminal	60 (12 people/km^2)	approx. 350 (50 people/km^2)	projection for 1990: 14 (2.2 people/km^2)	0 but recreation area within distance
Number of people living within 5 km of terminal	858 (28.9 people/km^2)	approx. 8000 (200 people/km^2)	projection for 1990: 98 (2.5 people/km^2)	5900 (151 people/km^2)
Number of people living within 10 km of terminal	9800 (85 people/km^2)	approx. 100 000 (470 people/km^2)	data from year 1977: 129 (0.9 people/km^2)	43 000 (275 people/km^2)

(ii) *Kinematic modeling.* In SAI-USA ship collisions are analyzed by assuming ship motions to be random in a zone of interest corresponding to the short interval of time preceding an accident. A kinematic model provides the expected number of collisions per year under this assumption for a harbor with specific characteristics of configuration and traffic. A calibration to the actual average conditions of seven harbors is then made by scaling the model to fit actual past collision frequencies in these harbors.

The estimation of the probabilities of various spill sizes due to the six different types of events considered in the reports are given in Table 3. It should be mentioned that the estimates given in Table 3 are not always taken directly from the reports. In some cases the estimates were adjusted to take additional data into account. SAI-USA used more ships with larger tanks than currently planned, so the probabilities and spill sizes were reduced accordingly. FERC-USA only considered spill sizes of 25 000 m^3 in their report, although they stated the data for smaller sizes as well. These data were considered in generating Table 3. KRAPP1,2,3-D produced a variety of different results using different accident reduction factors, ranging from 1.0 to 0.05. Because the latter factor was not based on any stated reasoning, we used the factor 1.0, which was used in KRAPP1-D.

The most interesting findings from this comparison of assessments were:

(a) compared with the probability of collision, grounding and ramming, the other events are rather unlikely (except for the internal failure in ACTION-UK);
(b) the differences in probabilities of spills between the three reports for Point Conception are substantial (between 10^{-3} and 10^{-6} for 10 000-25 000 m^3 spills);
(c) although the traffic patterns at Eemshaven, Mossmorran, and Wilhelmshaven are quite different, they all come up with a total spill probability of the order of 10^{-3}, but spill sizes differ and are not defined for Wilhelmshaven.

The event that could create the largest spill is the rupture of a storage tank. In the literature, it is assumed that one of the following events can cause a rupture: severe winds, airplane and missile crash, meteorites, earthquakes, internal system failure, and accidents at other chemical plants nearby.

The estimate of TNO1-NL is taken from historical data of a peak-shaving LNG plant. CREM-UK only qualify the probability as "remote", without reference to how this qualification was produced. ADL2-USA and SAI-USA derive their estimates from historical data on weather conditions, earthquake frequencies, and frequencies of airplane crashes. The probabilities for internal system failure—due to matallurgical failures—were derived through a technical analysis, considering the material and the variations of the temperature of the material causing fatigue or stress. BROTZ-D estimates the probability of an airplane hitting one of the six tanks from historical data from the FRG.

All storage tanks are placed within containment basins capable of holding all the contents (in liquefied form) of the tanks. All credible failure scenarios assume that these containment basins will not break and therefore all spills remain within these basins.

Only SAI-USA considers probability of rupture of more than one tank at a time, due to a common cause. The maximum credible spill is then considered as a rupture of all three storage tanks (each consisting of 77 500 m^3) at a time. SAI-USA adjust their probabilities to the fact that the tanks are empty approximately 40 percent of the time.

Major findings on storage tank rupture probabilities are:

(a) The probability of a storage tank rupture is estimated for all sites (except Mossmorran and possibly Wilhelmshaven, where not all reports are available for comment) of being of the order of 10^{-5} per year.

(b) As a conservative assumption the spill size is generally assumed to be at least the complete contents of one tank. However, CREM-UK only assume 15 percent of the contents of one storage tank to be spilled.
(c) There are no major differences in the estimates, except between ADL2-USA and SAI-USA. For example, the SAI probability of a spill due to objects crashing into the tank is 4×10^{-7}, while the ADL estimate is 10^{-5}. Elizabeth Drake (of ADL) has pointed out that this difference is due to changes in missile launch plans at the nearby Vandenburg Air Force base between the times the two reports were written (personal communication 1981).
(d) Common-cause failures causing more than one tank to rupture are only considered by SAI-USA.

3.2. Physical Consequences of LEG Spills

We have so far discussed the probabilities of different spill sizes resulting from failures of parts of the system. Before we can quantify the number of fatalities certain spill sizes could cause, we have to discusss what happens to the spilled LEG and how it can cause fatalities.

There seems to be agreement that only ignition and consequent rapid burning or detonation of the spilled LEG can have consequences to life and limb, because of thermal radiation and blast effects. LEG will immediately start to vaporize after a spill, resulting in a vapor cloud. This vapor cloud will then travel downwind and disperse. If there is no ignition, all parts of the cloud will eventually reach the lower flammability limit of concentration, below which it cannot be ignited. To estimate the effects it is therefore necessary to estimate the size of the vapor cloud, the downwind travel distance of the part of the cloud that retains a concentration above its lower flammability limit and the probability of ignition.

We will first discuss the size of the vapor cloud, which depends on the spill size, on meteorological conditions, and on whether the spill is on land or on water. We will then discuss estimates for the ignition probabilities at different sites and for different events.

3.2.1. Vaporization and Dispersion of LEG After a Spill

Among all topics of LEG risk assessment the question of how LEG behaves after a spill has attracted the most scientific interest. So far, empirical data include only information for spills up to 50 m^3 for an LNG spill on land, and up to 200 m^3 for an LNG spill in water. The prediction of the behavior of large spills has therefore had to rely on theoretical models, which are not easy to validate. Predictions differ for large spills but produce good estimates of the observed spills. The predicted downwind distances after a spill at sea, taken from the different reports, are listed in Table 4. It should be noted that these predictions are valid primarily over water, where the landscape does not influence vapor cloud dispersion in a specific way. One could expect that vapor cloud dispersion is faster over rough landscape, except in the case of propane and butane vapor, which could accumulate in low areas due to their high density.

The differences between the reports are substantial. While SAI-USA and BROTZ-D predict relatively short distances, ADL2-USA and FERC-USA are comparable in their prediction of large distances. It is also worth noting that the distance increases with decreasing wind speed in FERC-USA while for SAI-USA the distance decreases with decreasing wind speed.

Although possibly larger in size, spills on land are generally considered less dangerous than spills on water. The first reason for this assumption is that spills on land are confined because the storage tanks are surrounded by dikes, which are generally considered not to rupture. The second reason is that the vaporization rate of LEG on land is slower than on water.

Table 3. Estimation of LEG vessel failures.

	TNO1-NL	ACTION-UK	ADL2-USA	FERC-USA	SAI-USA	BROTZ-D	KRAPP1,2,3-D
(1) Probability of collision that can lead to a spill per ship approaching the LEG terminal	$2.8x10^{-5}$			$5x10^{-4}$	$1.3x10^{-8}$	—	$4x10^{-5}$
(2) Probability of grounding that can lead to a spill per ship approaching the LEG terminal	$2.5x10^{-4}$	$1.5x10^{-5}$ includes (2) and (3)		$4x10^{-4}$	0	—	$7x10^{-5}$
(3) Probability of ramming that can lead to a spill per ship approaching the LEG terminal	—		See (14)	$3x10^{-4}$	0	—	$3x10^{-7}$
(4) Probability of missile or airplane crash causing one spill per year	—	—		—	$4x10^{-7}$	$8.3x10^{-5}$	—
(5) Probability per year of a meteorite falling on a specific area of 1 m^2	—	—		—	$3.3x10^{-13}$	—	—
(6) Probability of internal system failure	—	$3.2x10^{-3}$		—	$1.0x10^{-11}$	—	—
(7) Number of ships per year	54	80	190	190	190	432	432
(8) Deck-size of ship in m^2 (maximum)	12 000	6600	12 000	12 000	12 000	12 000	12 000
(9) Length of stay of loaded ship in the vicinity of the terminal (years)	—	—	$2x10^{-3}$	$2x10^{-3}$	$2x10^{-3}$	$2x10^{-3}$	$2x10^{-3}$
(10) Size of one tank (maximum) in m^3	25 000	12 000	25 000	25 000	25 000	25 000	25 000

Table 3. Continued.

	TNO1-NL	ACTION-UK	ADL2-USA	FERC-USA	SAI-USA	BROTZ-D	KRAPP1,2,3-D
(11) Probability of different spill sizes given (1)							
0< ≤ 1 000 m^3	0	0		0.02	0		
1 000< ≤10 000 m^3	0	0		0.026	0		0.05
10 000< ≤25 000 m^3	0.56	0.25	See (14)	$2.3x10^{-2}$	0.22	—	spill size
25 000< ≤50 000 m^3	0.44	0		0	0.025		not defined
50 000< ≤75 000 m^3	0	0		0	0		
(12) Probability of different spill sizes given (2)							
0< ≤ 1 000 m^3	0	—		0.0024	—		
1 000< ≤10 000 m^3	0.33	—		0.0057	—		0.009
10 000< ≤25 000 m^3	0	—		$3.9x10^{-3}$	—	—	spill size
25 000< ≤50 000 m^3	0	—		0	—		not defined
50 000< ≤75 000 m^3	0	—		0	—		
(13) Probability of different spill sizes given (3)							
0< ≤ 1 000 m^3	—	—		0.0034	—		
1 000< ≤10 000 m^3	—	—	See	0.0065	—		0.1
10 000< ≤25 000 m^3	—	—	(14)	0	—	—	spill size
25 000< ≤50 000 m^3	—	—		0	—		not defined
50 000< ≤75 000 m^3	—	—		0	—		
(14) Total probability of different spill sizes per year*							
0< ≤ 1 000 m^3	0	0	0	$2.3x10^{-3}$	0		
1 000< ≤10 000 m^3	$4.5x10^{-3}$	$1.1x10^{-3}$	0	$3.3x10^{-3}$	0		$3.8x10^{-3}$
10 000< ≤25 000 m^3	$8x10^{-4}$	0	$7.4x10^{-5}$	$2.5x10^{-3}$	$8.9x10^{-7}$	—	spill size
25 000< ≤50 000 m^3	$7x10^{-4}$	0	$3.2x10^{-6}$	0	$9.9x10^{-8}$		not defined
50 000< ≤75 000 m^3	0	0	$6.5x10^{-9}$	0	0		

* = [(1)(11) + (2)(12) + (3)(13) + (5)(8)(9)] (7) + (4) + (6)

Table 4. Maximum downwind distance of a flammable vapor cloud following an instantaneous spill of LNG onto water.

Report	LEG spill size (m^3)	Atmospheric stability	Wind speed (km/h)	Downwind distance (km)
BROTZ-D	20 000	A-F	All wind speeds	2.3
		During night only	All wind speeds	3.5
TNO1-NL	25 000	D	—	3.3
		E,F	—	10.0
ADL2-USA	25 000	A	25.0	1.0
		D	21.0	7.0
		E	19.8	10.0
		F	10.8	20.0
FERC-USA	30 000	A	25.0	0.5
			16.0	0.5
			9.0	0.6
		D	25.0	4.2
			16.0	4.9
			9.0	5.9
		E	25.0	7.8
			16.0	9.2
			9.0	11.3
		F	25.0	18.1
			16.0	21.6
			9.0	27.1
SAI-USA	37 500	A,D,F	54.0	6.0
		A,D,F	25.0	3.5
		A,D,F	11.0	2.0
		A,D,F	0.0	1.0
ADL2-USA	50 000	A	25,0	1.0
		D	21.0	9.0
		E	19.8	15.0
		F	10.8	25.0
SAI-USA	88 000	A,D,F	11.0	2.5

3.2.2. Ignition of Vapor Clouds

Ignition probability is composed of two parts. The first is the direct ignition by the event that caused the spill. As can be seen from Table 5 these probabilities, depending on the different events, are generally high because it is assumed that an event that causes a tank to rupture could also create enough frictional heat to ignite the resulting vapor cloud.

The second part is the probability that the vapor cloud is ignited by some other source given that it is not ignited immediately. Obviously this depends on the availability of ignition sources within the flammable bounds of the vapor cloud. Delayed ignition will in general have larger consequences, because the vapor cloud increases in size and travels downwind. Therefore, for most spill locations a high immediate ignition probability will reduce the overall risk. In this respect TNO1-NL and ACTION-UK are more conservative in their estimates than

Table 5. Probabilities of immediate ignition following different events.

Event causing the ignition	TNO1-NL	ACTION-UK	FERC-USA	SAI-USA	BATTE2-OTH	KEEN-OTH
Vessel tank rupture caused by:						
collision	0.65	0.66	0.9	0.9	0.8	0.9-0.99
grounding	0.1	—	0.0	—	0.3	—
ramming	—	—	0.9	—	—	—
missile/airplane	—	—	—	0.9	0.9	—
meteorite	—	—	—	0.0	—	—
internal failure	—	0.9	—	0.0	—	—

the other reports. Certainly, the ignition probability can be site-dependent. For example, KEEN-OTH points out that the immediate ignition probability is estimated at a high value because collisions at the specific site studied would generally involve larger vessels carrying dangerous cargoes (such as chlorine). Because historical data on LNG spills are lacking, the estimated ignition probabilities can not be validated.

FERC-USA, SAI-USA, BATTE2-OTH, and KEEN-OTH use the same model for delayed ignition probability. They assume that each source of ignition has the same probability p of igniting the vapor cloud. Thus the probability P_n that the vapor cloud will have been ignited within n seconds becomes $P_n = 1 - (1 - p)^n$. Additionally, all assessments using this model assume that each person (or that every fourth person) is a source of ignition, because (s)he will use facilities (e.g., car, oven, light) that are actual sources of ignition. The differences between the reports are the judgmental estimates of the probability p (Table 6).

Table 6. Ignition probabilities per person in case of delayed ignition.

	FERC-USA	SAI-USA	BATTE2-OTH	KEEN-OTH
Probability p that each person within the vapor cloud ignites the cloud	0.0025	0.1	0.01	0.01-0.1

Any of the assumed values of p can be either conservative or nonconservative depending on the number of people (and thus ignition sources) within the reach of the vapor cloud. The estimate of FERC-USA, for example, is less conservative for Point Conception than the estimate of SAI-USA because there are not more than 130 people living within 10 km of the LNG facility. Thus the FERC-USA estimate implies that there is a substantial probability that the vapor cloud will not be ignited at all, while the estimate of SAI-USA implies that the vapor cloud will be ignited with very high probability. On the other hand, using the model for Wilhelmshaven with 43 000 people living within 10 km of the LNG site, the FERC-USA estimate implies that the vapor cloud will be ignited, but only after covering more populated area than that predicted using the SAI-USA estimate.

3.2.3. *Fatalities Caused by Ignited Vapor Clouds*

Effects from ignited vapor clouds can be twofold: thermal and blast effects. There is no doubt that thermal effects exist, but it is an open question whether blast effects due to a deflagration or detonation can occur at all with methane and, if so, whether the peak overpressure created by a deflagration or detonation will be significant enough to cause damage. TNO1-NL considers blast effects to be the only serious danger, and thermal effects are of comparatively minor importance. CREM-UK considers both thermal and blast effects, as is logical since the Mossmorran terminal handles butane, propane, and ethylene, which are known to explode in certain mixtures with air. ADL2-USA only considers thermal effects, because an explosion (either deflagration or detonation) of methane is considered very unlikely. FERC-USA and SAI-USA again only consider thermal effects. BROTZ-D considers both thermal and blast effects. In NMAB-REV it is concluded that the possibility of explosions of LNG vapor clouds cannot be ruled out completely, although empirical evidence for such a possibility does not exist.

One first step to estimate the percentage of fatalities within certain distances from the vapor cloud is to state the level of thermal radiation and peak overpressure above which fatalities can be expected. Here one has to distinguish between primary and secondary effects. Primary effects are fatalities directly caused by thermal radiation and peak overpressure; while secondary effects are fatalities caused by fires created from thermal radiation by collapsing buildings as a result of peak overpressure.

All reports available to us consider only primary thermal and secondary blast effects. BROTZ-D maintains that primary blast effects can be ruled out, because the required peak overpressure has never been observed. Secondary thermal effects, however, are a possibility for people sheltered from direct radiation, but are very difficult to estimate. One way to include secondary thermal effects is to assume a low radiation level as a threshold level for fatalities.

Blast effects do not play a significant role in the risk calculations in most of the assessment studied. The only report relating blast effects to fatalities is TNO1-NL, and BROTZ-D does not consider them at all. The treatment of thermal effects varies markedly among the assessments. The distance from the center of the fire to the lower fatality level is about twice as large in ADL2-USA as in FERC-USA and SAI-USA. CREM-UK and BROTZ-D do not give a lower fatality level.

The major findings of a comparison of fatality calculations among the assessments can be summarized as follows:

(a) The reports differ on the major cause of fatalities. While TNO1-NL assumes all fatalities to be caused by secondary effects of vapor cloud explosions, ADL2-USA, FERC-USA, and SAI-USA assume fatalities to be caused by thermal radiation. CREM-UK and BROTZ-D do not consider fatalities as a result of ignited vapor clouds.

(b) There is also some difference in the radiation levels above which there will be fatalities. ADL2-USA adopts the most conservative assumptions on this topic among the reports.

(c) The effects of LNG and LPG vapor clouds can be quite different. For example, it is known that LPG vapor clouds can explode, while the possibility of an unconfined LNG vapor explosion has not yet been determined.

(d) The ignition of an LNG vapor cloud can have effects on nearby plants with possibly high secondary effects on the people living or working near the plants. Except at Point Conception there are chemical plants near all the other LEG terminals. CREM-UK and BROTZ-D considered this point and concluded that effects on the chemical plants nearby do not increase the overall risk significantly. In TNO1-NL it is pointed out that in the case of a detonation a nearby NH_3 storage tank could collapse with inadmissible consequences (the lethal dose of NH_3 would reach tens of kilometers).

3.3. Assessment of Population Risk

In many of the assessments the various numbers discussed so far are combined to aggregate estimates of population risk. These estimates of the societal risk, the individual risk and the risk of multiple fatalities are given in Table 7. No estimates of the risks were given in CREM-UK and BROTZ-D.

Not surprisingly, Point Conception has the lowest risk among the three sites. However, as discussed above, different reports consider quite different events. The probabilities also vary for the same event and the same site between different reports. It should also be noted that the estimate of SAI-USA was given for an LNG terminal with more storage tanks and larger ships than the one currently planned. Although we adjusted the estimates in earlier sections accordingly to make them comparable with ADL2-USA and FERC-USA, this was not done in Table 7. Therefore, the risk of the smaller LNG terminal currently planned, as estimated by the SAI-USA analysis, would be lower than that stated in Table 7.

Table 7. Estimates of risks for the different sites.

	TNO1-NL	ACTION-UK	ADL2-USA	FERC-USA	SAI-USA
Societal risk (fatalities per year)	$4x10^{-2}$	—	$7x10^{-6}$	$1x10^{-5}$	$1x10^{-6}$
Individual risk (probability of fatality per year)	$\leq 7x10^{-6}$	$7x10^{-4}$	$\leq 9x10^{-8}$	$8x10^{-7}$	$1x10^{-8}$
Number of people at risk	$\geq$5000	?	$\geq$80	15	90
Risk of multiple fatalities: probability that number of fatalities per year is equal to or greater than					
1	$3x10^{-3}$	—	$1x10^{-6}$	—	$6x10^{-7}$
10	$1x10^{-3}$	—	$1x10^{-8}$ $6x10^{-7}$	—	$3x10^{-11}$
100	$5x10^{-6}$	—	ø	—	ø
1000	$5x10^{-6}$	—	ø	—	ø
5000	$3x10^{-7}$	—	ø	—	ø

4. CONCLUSIONS

The major findings of this report can be summarized as follows.

(a) There is no unique concept of risk that is used throughout all the risk assessment reports examined in this study. Many of the important differences between the reports stem from the different risk concepts used. Some reports do not even define their underlying risk concepts. However, there is a concept of risk that involves several measures, each based on both probabilities of failures and consequences of failures, that we judge to be superior to other less comprehensive risk concepts, and that we have adopted in this study.

(b) The possible failures of the system, the probability of those failures, and the estimation of their consequences to life and limb differ between the reports. Not all the differences can be explained by differences between the terminals and sites; some must be viewed as resulting from the limited knowledge and understanding of LEG risks. In this respect too little reference is made to remaining uncertainties in the estimation of risk in most reports.

(c) Given the differences between the reports there is no relative tendency for each report individually to over- or underestimate the risk. Rather, each report is more conservative on certain topics and less so on others, as compared with the other reports. Thus no report can be singled out as producing a more conservative estimate of the risk (with respect to all parts of the total risk) than any of the others.

(d) On a relative risk scale it can be said that, of the four sites, Point Conception presents the lowest societal risk (because of very low population density), Mossmorran and Wilhelmshaven present the highest relative risk (because of high population density and more vessel traffic) and Eemshaven is in between.

(e) Although risk is an important dimension of the decision to import LEG and to choose a specific site for the terminal, it should not be forgotten that other dimensions such as reliability are important too. Any decision regarding LEG importation and terminal siting should involve comparisons with alternative options. As part of that process the risk of LEG should be compared with the risk of other options.

(f) Whatever flaws the LEG risk assessments may have, they are clearly superior to less systematic ways of identifying possible system weaknesses and informing the decision-making process on the topic of risk.

ACKNOWLEDGMENTS

We would like to acknowledge the assistance of the many people who made helpful comments on earlier drafts of this paper. The accuracy of a review with the breadth of this one depends on the critical responses of the community of analysts reviewed. While the people who commented are too numerous to mention here, we would like to acknowledge in particular the help of Robert Arvedlund, Arthur Boni, Elisabeth Drake, William Fairley, Ralph Keeney, Jerry Kopecek, H.J. Nikodem, Elisabeth Pâte-Cornell, and Lloyd Philipson. While we acknowledge the assistance of these experts, we accept full responsibility for the test and such errors as remain.

The research reported in this paper is supported by the Bundesministerium für Forschung und Technologie, FRG, contract No. 321/7591/RGB 8001. While support for this work is gratefully acknowledged, the views expressed are the author's own and are not necessarily shared by the sponsor.

REFERENCES

Risk Assessment Reports for Eemshaven (NL)

TNO1-NL: van Amerongen, H., Bouma, J.W.J., and Snellink, G. (1978) Evaluatie va de gevaren verbonden aan aanvoer, overslag en opslag van vloeibaar aardgas met betrekking tot een Eemshaven-terminal. Rijswijk, Netherlands: TNO Bureau Industriele Veiligheid.

Risk Assessment Reports for Mossmorran (UK)

ACTION-UK: (1979) Mossmorran-Braefoot Bay: Shipping Hazards. Aberdour, UK: Aberdour and Dalgety Bay Joint Action Group.

CREM-UK: Cremer and Warner (1977) The Hazard and Environmental Impact of the Proposed Shell LNG Plant and Esso Ethylene Plant at Mossmorran, and Export

Facilities at Braefoot Bay, Vols. I and II. A Report prepared for Fife Regional Council, Dunfermline District Council and Kircaldy District Council.

Risk Assessment Reports for Point Conception (USA)

ADL1-USA: Arthur D. Little, Inc. (1978a) Draft Environmental Impact Report for Proposed Point Conception LNG Project, prepared for California Public Utilities Commission. Cambridge, Mass.: Arthur D. Little, Inc.

ADL2-USA: Arthur D. Little, Inc. (1978b) LNG Safety. Technical Report 16 of Draft Environmental Impact Report for Proposed Point Conception LNG Project. Cambridge, Mass.: Arthur D. Little, Inc.

FERC-USA: Federal Energy Regulatory Commission (1978) Final Environmental Impact Statement, Vol. III Comments and Appendixes, Western LNG Project FERC/EIS-0002F. Washington, DC: FERC.

SAI-USA: Science Applications, Inc. (1976) Terminal Risk Assessment Study for Point Conception, California, prepared for Western LNG Terminal Company. La Jolla, California: Science Applications, Inc.

Risk Assessment Reports for Wilhelmshaven (D)

BROTZ-D: Brötz, W. (1978) Sicherheitstechnisches Gutachten zum Planfeststellungverfahren eines Schiffsanlegers vor dem nördlichen Teil des Voslapper Grodens im Norden von Wilhelmshaven. Stuttgart: Institut für Technische Chemie der Universität Stuttgart.

KRAPP1-D: Krappinger, O. (1978) Abschätzung des Risikos, das Tanker mit gefährlicher Ladung im Jadefahrwasser mit anderen Schiffen kollidieren oder auf Grund laufen. Hamburg: Hamburgische Schiffbau-Versuchsanstalt GmbH.

KRAPP2-D: Krappinger, O. (1978) Risikoanalyse: Über die Gefährdung der an den Umschlagebrücken der DFTG und ICI liegenden Schiffe durch den die Anlage passierenden Verkehr auf dem Fahrwasser der Jade. Hamburg: Hamburgische Schiffbau-Versuchsanstalt GmbH.

KRAPP3-D: Krappinger, O. (1978) Ergänzung der Risikoanalyse: Über die Gefährdung der an den Umschlagsbrücken der DFTG und ICI liegenden Schiffe durch den die Anlage passierenden Verkehr auf dem Fahrwasser der Jade. Hamburg: Hamburgische Schiffbau-Versuchsanstalt GmbH.

WSD-D: (1978) Bericht Transport gefährlicher Güter zu den geplanten DFTG/ICI Umschlagbrücken. Aurich: Wasser- und Schiffahrtsdirektion Nordwest.

Other Risk Assessment Reports (OTH)

BATTE1-OTH: (1978) Risk Assessment Study for the Harbor of Gothenburg, prepared for the Swedish Energy Commission. Franfurt: Battelle-Institut.

BATTE2-OTH: (1978) Risk Assessment Study for an Assumed LNG terminal in the Lysekil Area, prepared for the Swedish Energy Commission. Frankfurt: Battelle Institut.

HSC-OTH: (1978) Canvey—An investigation of Potential Hazards from Operations in the Canvey Island/Thurrock Area. London: Health and Safety Executive.

KEEN-OTH: Keeney, R., Kulkarni, R., and Nair, K. (1979) A risk analysis of an LNG terminal. Omega 7:191-205.

SES-OTH: (1976) Environmental Impact Report for the Proposed Oxnard LNG Facilities. Draft EIR Appendix B. Los Angeles, California: Socio-Economic Systems, Inc.

Reviews of Risk Assessments (REV)

HAZEL-REV: Hazelwood, R.N., and Philipson, L.L. (1977) Survey of LNG Risk Assessment, prepared for California Public Utilities Commission. Los Angeles, California: Socio-Economic Systems, Inc.

LEWIS-REV: Lewis, H.W. (Chairman)(1978) Risk Assessment Review Group Report to the US Nuclear Regulatory Commission, NUREG/CR-0400. Washington, DC: US Nuclear Regulatory Commission.

NMAB-REV: (1980) Safety Aspects of Liquefied Natural Gas in the Marine Environment, NMAB 354. Washington, DC: National Academy of Sciences, National Materials Advisory Board.

General Papers (GEN)

LUCE-GEN: Luce, R.D., and Raiffa, H. (1957) Games and Decisions. New York: Wiley.

PHIL-GEN: Philipson, L.L. (1978) Safety of LNG systems. Energy Systems 4:135-55.

Societal Response to Three Mile Island and the Kemeny Commission Report

Roger E. Kasperson and Arnold Gray

Center for Technology, Environment and Development, Clark University, Worcester, Massachusetts 01610, USA

1. INTRODUCTION

The accident at the Three Mile Island nuclear power plant on 28 March 1979 was by common consensus the worst to occur in the history of commercial nuclear power generation in the United States. It is not surprising, therefore, that the accident provoked a series of assessments of its meaning as to the safety of nuclear power. Prominent among these was the President's Commission on the Accident at Three Mile Island, popularly known as the Kemeny Commission.

The commission labored for six months, eventually taking some 150 formal depositions, interviewing hundreds of individuals, hearing testimony under oath from numerous witnesses, and collecting sufficient material to fill 300 feet of library shelf space. In its work, the commission was supported by a budget of $1 million and a substantial staff. Its final report, issued in October 1979, received undoubtedly more media coverage and congressional attention than any other document on nuclear power safety.

The report is one of a genre of risk assessments. Unlike the Reactor Safety Study (WASH 1400) or the Risk Assessment Review Group Report (the Lewis Report) which relied heavily upon expert assessment dealing with the quantitative probabilistic assessment of risk, the Kemeny Commission inquired into the larger issues of nuclear safety as indicated by a particular accident. Because of the significance of the crisis event and the direct responsibility of the Commission to the President, the report had a unique opportunity to contribute to the shaping of nuclear safety in the United States.

This paper inquires into the response of various segments of society, particularly the mass media, industry, and the regulators.

2. A NOTE ON METHODOLOGICAL ISSUES

Assessing the impact of the Kemeny Commission Report on nuclear safety policy requires the isolation of the report from the numerous other risk assessments conducted after the accident, from the accident itself, and from the ten congressional subcommittees that had held hearings on the subject by the first anniversay of the Three Mile Island (TMI) accident. This cannot be done. In fact, the Nuclear Regulatory Commission (NRC) quite explicitly and systematically integrated the various report findings in order to fashion a coordinated response. In addition, a number of safety problems were quite evident in the accident itself, and it is futile to determine which source stimulated a particular response.

Within these constraints, however, there are some opportunities. A substantial part of the industry and governmental response occurred well in advance of the publication of the Kemeny Report some seven months after the accident; more of it presumably would have occurred even in the absence of the commission. Also, several post-accident evaluations and congressional inquiries appeared prior to the report and thus provide a benchmark from which to assess its particular contributions. Finally, there was not a complete overlap in these reports so that some of the individual findings and recommendations of the Kemeny Report can be distinguished and assessed as to impacts.

3. NEWSPAPER COVERAGE OF THE KEMENY REPORT

In regard to media coverage, the Kemeny Commission Report represents a special case among nuclear risk assessments. First, it came in the wake of the TMI accident, the top news story of the year. Consequently, the commission operated under a spotlight of media attention from the time it was created by President Carter until it published its report some seven months later. Second, unlike the Rasmussen and Lewis Reports, the Kemeny Report was intended to investigate the problems that led to an *actual* event and then to make recommendations on how best to avoid any similar occurrences in the future.

3.1. Pre-report Coverage

Interest in the Kemeny Commission was intense well before the report actually appeared. A total of 31 articles and three editorials on the Commission's activity were published prior to its release on 31 October, more than the total number of articles listed in the *New York Times Index* after its release. This well developed interest in the commission's investigatory work is apparent from Figure 1. Since further references to the Kemeny Report were included in articles on related topics, (e.g., the Rogovin Commission) after publication, the column inch counts suggest that pre- and post-Kemeny coverage was roughly equal.

The creation of the Kemeny Commission was first reported in the *New York Times* on 6 April 1979, after President Carter's announcement that a presidential commission of experts would be convened in order to "investigate the causes of this accident and ... make recommendations on how we can improve the safety of nuclear power plants". The first articles on the commission that tended to "paint a picture" of the TMI accident, and nuclear power in general, appeared on 20 May. The commission had just held its first day hearing testimony from residents of Middletown, and the article emphasized the emotionalism of the session: "Citing estimates that a few additional cases of cancer might develop as a result of the accident... a resident asked, his voice rising with emotion: 'Who'll be the ones? Myself? My son? My wife?'" As the testimonies were given, first with the control room operators, then with NRC officials, Babcock and Wilcox, executives, and Pennsylvania state officials, an image of mismanagement, carelessness, ineptitude, and complacency emerged. The reader of the *New York Times* could not easily come away from these articles with anything but a generally pessimistic view of nuclear power in America.

With the evidence in and the jury in deliberation, the press was left to wait for the verdict. Following up on rumors and strategically placed leaks, the *New York Times* focused much of its attention upon an anticipated moratorium of some form. Apart from this issue, the preliminary findings alluded to in the press did, for the most part, show up in the actual Kemeny Report. Key among them were:

(i) The NRC had a major attitudinal problem and was preoccupied with licensing. It would be recommended that the NRC be reorganized as an executive agency.
(ii) There must be an approved licensing plan.
(iii) There should be periodic relicensing of nuclear power plants.
(iv) Operator training should be upgraded with increased government regulation and better (possibly standardized) design for control rooms.

None of these findings or recommendations, however, received the same amount of attention that the moratorium issue received.

3.2. Post-report Coverage

The Kemeny Report ultimately included 81 specific findings and 44 recommendations. Of the 81 findings, only 13 were reported in the *New York Times* (see Table 1). Of the 44 recommendations, only eight were covered. The treatment, in short,

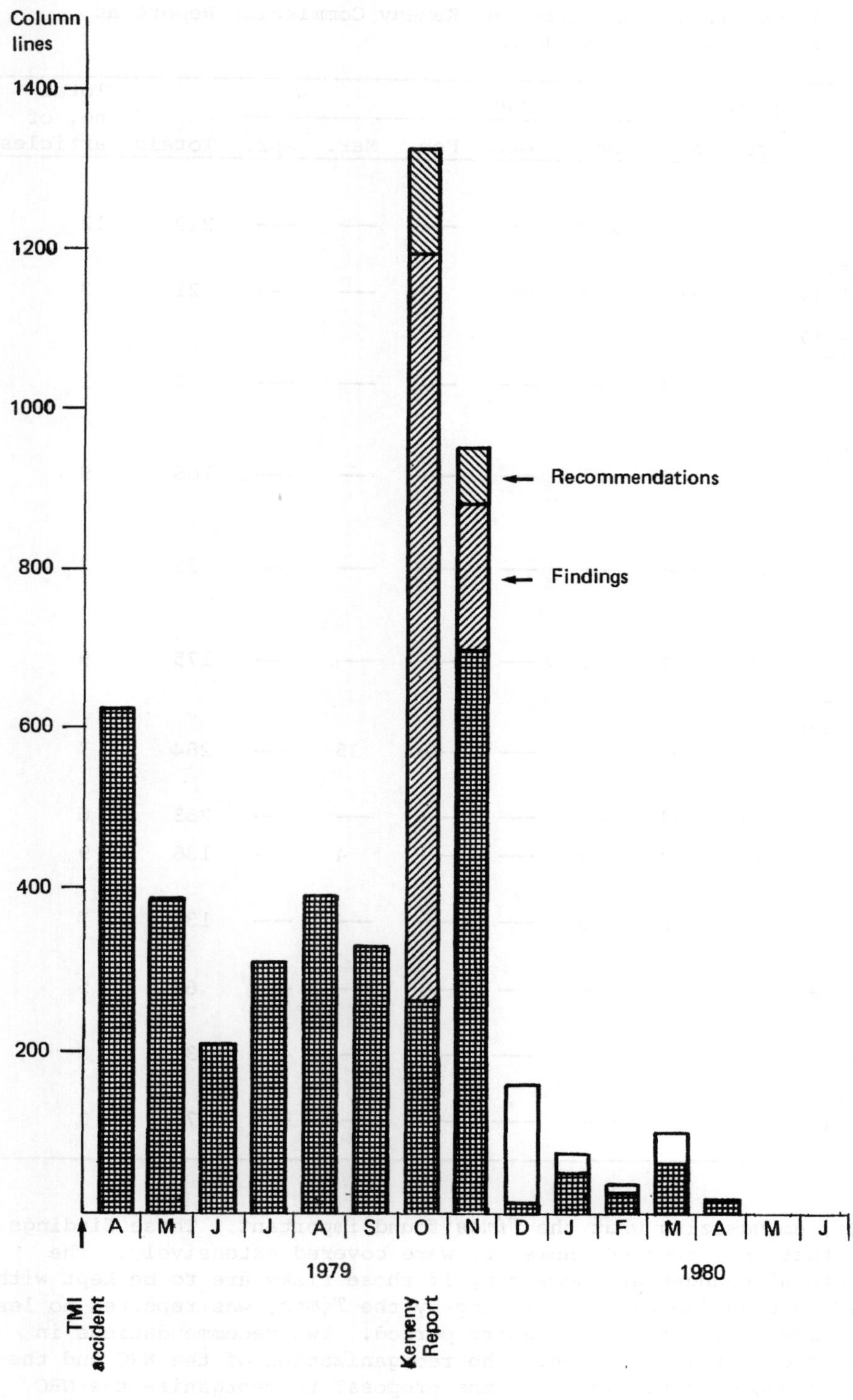

Figure 1. Column lines of Kemeny Commission general coverage, findings, and recommendations in the *New York Times*.

Table 1. Findings and recommendations from the Kemeny Commission Report as reported in the *New York Times* (in print lines).

	1979			1980					Total no. of
	Oct.	Nov.	Dec.	Jan.	Feb.	Mar.	Apr.	Totals	articles
Recommendations									
A1—Reorganize NRC	70	30	119	9	—	—	—	212	13
A2—Establish executive oversight committee	6	—	8	—	7	—	—	21	3
A5, F1—Broaden utility responsibility in emergency	—	3	—	—	—	—	—	3	1
A7, A8, A11e, F1—Upgrade licensing procedure	50	30	21	—	—	—	—	106	9
C1—Improve operator and operating procedure	22	—	—	—	—	—	—	22	3
Findings									
Overall conclusion	136	24	15	—	—	—	—	175	9
A1, 2, 8, 12—assessment of significant events	249	—	—	—	—	35	—	284	7
H4, 10—Public right to know	159	104	—	—	—	—	—	263	6
G8, 12—NRC	117	12	3	—	—	4	—	136	9
E5—Utility and supplier	107	20	3	—	—	—	—	130	7
B16, 4—Public health and health effects	54	14	—	—	—	—	—	68	5
F1—Operators and operating personnel	38	—	—	—	—	—	—	38	3
Attitudes: personal and institutional	78	—	—	—	—	—	—	78	3

was highly selective, emphasizing what the *Times* found important. Those findings and recommendations that were covered, however, were covered extensively. The finding that "fundamental changes are necessary if those risks are to be kept within tolerable limits", deemed *the* central finding by the *Times*, was reported no less than nine separate times during the post-report period. Two recommendations in particular dominated the overall coverage: the reorganization of the NRC and the upgrading of licensing procedures. In fact, the proposal to reorganize the NRC (not adopted) received more coverage than the remaining 40 or 50 recommendations. Even the recommended upgrading of licensing procedures was discussed primarily in terms of the moratorium issue—an issue, we should note, that while not a recommendation, received more coverage than any of the actual Kemeny recommendations or findings.

Coverage of the report rapidly fell off the *Times* reporting agenda, although analysis of the coverage is difficult because attention shifted to specific issues rather than the report itself. A gleaning of the *Times Index* suggests that attention shifted toward transient conditions at various nuclear plants, problems surrounding Indian Point, economic problems of the nuclear industry (including fines, cost over-runs, delays), and the political struggles between pro- and anti-nuclear forces rather than the important but less conspicuous responses within industry and government dealing with the generic safety issues raised by TMI. Throughout 1980, for example, there was no substantial discussion of the NRC's *Action Plan* (NUREG 01600), the single most important governmental response to emerge from the accident and the Kemeny Commission Report.

4. SCIENTIFIC PRESS COVERAGE

The coverage of the Kemeny Report in *Science* and *Nature* was quite different from that of the *New York Times* although, in several respects *Science* and *Nature* also differed from one another. *Science* carried three articles on the Kemeny Commission before the report release, treating the make-up of the committee and its budget and constraints, the ending of the licensing moratorium, and the iodine-131 problems. *Nature* mentioned the Kemeny Commission only twice prior to the report and referred to completely different issues: the California study of TMI, and a news brief on the dissolution of the citizens' panel.

Both *Science* and *Nature* published only one article to cover the report's findings and recommendations. Both provided extensive coverage of the report, although *Science* was somewhat more specific and comprehensive. Neither of the two, however, included a *verbatim* listing of either the findings or the recommendations. The *Science* article, under the heading "Kemeny Report: Abolish the NRC", referred to 23 specific findings, nine recommendations and two criticisms of omitted recommendations. The report was criticized for "not [going] the extra step and [demanding] fundamental changes" and also for not asking for a licensing moratorium.

The reporting in *Nature* was less complete, treating eight findings, five recommendations, plus a section on the happenings of the accident. *Nature* did go on, however, to discuss immediate reaction to the report by pro- and anti-nuclear groups. The critics had charged that the report's "bark may ... turn out worse than its bite". The nuclear industry interpreted the report's message as "proceed with caution". *Nature* later published an article (17 January 1980) that challenged the latter view.

In terms of post-Kemeny coverage, the two journals presented only one major article (*Science*, 21 December) specifically addressing responses to the report. Briefer treatments did appear as well. On 8 February 1980, *Science* dealt with the Rogovin Report, noting the agreement of the two reports on the need to reorganize the NRC. A second article covered the NRC's review of reactor design in view of the Crystal River accident. While the Kemeny Report was not mentioned *per se*, there was discussion of safety recommendations that had been included in the report.

Nature carried three post-Kemeny articles. The first covered the congressional debate over the moratorium issue, noting that the report did not recommend a licensing moratorium, and the second included a discussion of nuclear safety, with attacks authored by Dr Russell Peterson (a former Kemeny Commissioner). An alternative view was presented by Nobel prizewinner Dr Rosalyn Yalow. Finally, in an article (19 June) entitled "What (if any) future for nuclear power?", *Nature* criticized the Kemeny Report for its lack of criticism and its minimal impacts.

Overall, the scientific press, as indicated by coverage in *Nature* and *Science*, achieved a more balanced and analytical treatment than the *New York Times*, but again the follow-up coverage of responses was insufficient to provide the reader with an informed treatment of what the report eventually meant for nuclear power.

5. INSTITUTIONAL RESPONSES

To assess institutional responses to the accident and the Kemeny Report, we can identify 12 key areas of recommendation from among the 43 specific ones made by the commission. For each key recommendation, the major societal responses are noted, major unresolved issues specified, and our overall assessment of the response is provided in Table 2.

The Kemeny Commission reached a number of biting judgments concerning the primary institutions responsible for the assurance of nuclear safety, the most notable of which were as follows.

(i) *The Nuclear Regulatory Commission.* "with its present organization, staff, and attitudes, the NRC is unable to fulfil its responsibility for providing an acceptable level of safety for nuclear power plants" (p56).

(ii) *The Advisory Committee on Reactor Safeguards.* The Committee is the only body independent of the NRC staff which regularly reviews safety questions, but the Committee "has established no firm guidelines or procedures," its members are "part-time and have a very small staff" and it relies heavily upon the NRC staff for follow-up of concerns.

(iii) *The Utility.* The utility (Met Ed) failed in a number of important cases "to acquire enough information about safety problems, failed to analyze adequately what information they did require, or failed to act on that information" (p43). "It did not have sufficient knowledge, expertise, and personnel to operate the plant or maintain it adequately" (p44).

To deal with these institutional deficiencies, the commission recommended a broad set of changes involving the NRC, the Advisory Committee on Reactor Safeguards, and industry.

The Kemeny Commission found that the NRC lacked sufficient organizational and management capability to ensure safety, a judgment supported by the Rogovin Report. Unfortunately, the commission recommended the rather shopworn suggestion of agency reorganization, in this case a change from an independent regulatory commission to an executive branch agency with an administrator, as the most prominent means of redress. The Kemeny Report was the first accident post-mortem to call for this change, though it subsequently also found favor in the Rogovin Report. The recommendation was unpopular from the start: the NRC staff opposed it, all the current NRC commissioners save one also opposed it. Congress was lukewarm to the idea, and the President, sniffing congressional opposition, never supported the recommendation. Instead, he called for, and congress eventually approved, a strengthening of the chairman's role.

Two years after the accident, top leadership in the NRC remains as an outstanding problem, recently described by one of the NRC commissioners as "analogous to hitching fire horses at different points around a sled". The general weakening of reglatory agencies in the current Reagan administration does not bode well for the hope that the recent drift and indecision will halt and that coherent, effective leadership committed to safeguarding public health and the environment will emerge in the NRC. Other changes in the NRC have met with greater success and indicate some limited improvements in regulatory performance. Central to these responses has been a shift in commission resources and emphasis on monitoring and assessing operating reactors. Within four months of the accident (and well in advance of the Kemeny Report), the NRC established a new office for analysis and evaluation of operational data aimed at the serious deficiencies in learning from past reactor incidents and malfunctions apparent in the TMI accident. The NRC also established a program of resident inspectors stationed at individual power plants. The severity of licensing exams for reactor operators has also been increased, producing a rise in failure rates from 5 to 30 percent. The NRC has improved its capability for crisis management by clarifying responsibilities and improving communication with an analytical strength for existing reactors.

Table 2. Societal response to key Kemeny Commission recommendations (as of March 1981).

Recommendation	Response
Restructure/improve NRC (A1)	President does not accept Kemeny reorganization recommendations. Congress retains collegial structure with strengthened powers of chairman. Chairman designated as spokesman in emergencies.
	<u>Assessment:</u> basic problems of the commission referred to in report remain unresolved, restructuring is not achieved, but some improvement in emergency response and regulating of operating reactor capabilities. In September 1980, the Nuclear Safety Oversight Committee finds evidence of a "business-as-usual mindset in NRC".
Improve ACRS (A3)	NRC opposes any mandatory response to ACRS recommendations. On 11 February 1980, ACRS charges NRC "largely ignores" its input on Kemeny Commission responses.
	<u>Assessment:</u> no substantial action undertaken to improve ACRS. It is unlikely that the ACRS can and/or will influence change within the NRC.
Establish new oversight committee (A2)	Executive Order established Nuclear Safety Oversight Committee on 18 March 1980. Committee issues three letter reports to the President on NRC action plan, radiological consequences of nuclear accidents, and emergency response planning.
	<u>Assessment:</u> Committee has provided limited but useful function. Future is unclear.
Upgrade reactor operator and supervisor training (A4, C1, C4)	Nuclear Safety Analysis Center establishes computerized communication system connected to all utilities on operating incidents. NRC proposes upgrading in formal education: senior reactor operators, 60 college credits in engineering; shift supervisors, a BS degree in engineering. Utilities improve training for emergency events. No change proposed in formal education of reactor operators. Memphis State University inaugurates new training program in cooperation with utilities. Severity of licensing exams increased; failure rate rises from 5 to 30 percent. NRC declines to accredit training programs.
	<u>Assessment:</u> upgrading becoming evident though requirements still lag behind those in Europe.
Increase safety emphasis in licensing (A10)	NRC reorganizes licensing staff to correct weaknesses in licensing process. Increased attention to operator training, utility management, emergency planning, reactor design features, and evaluation of plant operating experience; NRC decides against Office of Hearing Counsel. 1981 licensing plan reduces role of intervenors.
	<u>Assessment:</u> actions to date fill a number of gaps in safety coverage, but the degree of substantial improvement unclear. NRC licensing of Sequoyah plant questions commitment to safety. Reduced role of intervenors weakens safety focus.

Table 2 continued.

Recommendation	Response
Improve safety inspection and enforcement (All)	NRC establishes resident inspectors at power plants, requires annual evaluation of licensees, improves reporting requirements. A new NRC Office for Analysis and Evaluation of Operational Data established (prior to Kemeny Report) in July 1979. Fines for utilities increased. Bingham Amendment calls for "systematic evaluation" of all operating nuclear power plants, a possible 5-8 year effort that has evoked opposition. Assessment: although too early to tell, indications are of substantial improvement in inspection and regulation of operating reactors. But position of top leadership of NRC during Reagan administration will be important. Bingham Amendment will require significant new NRC resources.
Improve technical assessment and equipment (D1-D3)	Utilities initiate improvement in control room design and instrumentation. Assessment: substantial improvements implemented or ongoing in improved instrumentation, equipment, and monitoring.
Initiate new reactor risk assessments (D4-5, D7, E1)	NRC reorientates risk assessment research program with new attention to higher probability events, accident mitigation, and human factors. Retrospective iodine release study of TMI accident suggests possible past overestimate of consequences by factor of 10. Utilities establish improved monitoring and dissemination system of operating incidents. NRC establishes Division of Human Factors and initiates effort to define level of acceptable risk. Epidemiological studies of effects of low-level radiation initiated. EPA recommends against ten-fold reduction in occupational standard. Probabilistic risk assessment initiated by utilities at eight power plants. Radiation Policy Council established in Executive Branch. Assessment: significant changes instituted to give new priority to TMI-like events, to human factors, and accident mitigation. Individual plant risk assessments should improve safety performance and enlarge accident response capability.
Improve industry attitudes and performance (B1-B3, B5)	Industry establishes two new institutions: Institute for Nuclear Power Operations (INPO) with power plant evaluation and training as primary functions and Nuclear Safety Analysis Center (NSAC) with analysis of operating experience and other technical assessment its primary activities. International cooperation with NSAC makes world experience database a possibility. Assessment: substantial industry response: new institutions are important safety vehicles. Still unresolved are prevailing attitudes and assurance of high level of overall technical competence in individual utility management structure.

Table 2 continued.

Recommendation	Response
More remote siting of nuclear power plants (A6)	NRC proposes (NUREG 0625) upper limits on population densities around plants and making siting criteria distinct from engineered safeguards. Estimates suggest 49 of 84 currently operating plants would fail to meet criteria. Strong industry opposition. Assessment: proposal currently mired in controversy; no change to date, but new plants not currently being ordered in any event. Since no retrospective application of criteria, limited safety impact on 100-150 GWE nuclear system.
Improve emergency response and mitigation (A7-A8, E3-E5, F1-F3, G1-F4)	NRC issues new rule on emergency response plans, extending 5-mile zone to 10-mile and 50-mile radii. All operating reactors required to have emergency plans approved by April 1981. NRC installs a crisis management communications link of all power plants to NRC headquarters. New rule mandates that state be able to notify every person within 10 miles of a nuclear power plant of accident within 15 minutes and evacuate population. Proposal to distribute potassium iodide pills mired in controversy. *Nucleonics Week* survey finds confused and uncertain response by states. No notable improvement in mass media capabilities, despite an NRC pilot program. Assessment: although utilities and the NRC have improved their emergency response capabilities, the overall capacity of society to respond to a major accident remains in doubt.
Educate the public (F4, G5)	NRC plans to investigate need for literature. No program instituted to date. Assessment: no substantive response despite widespread scientific belief as to need.

Beyond these useful changes, however, is the more basic and difficult problem of attitudes and orientations throughout the professional staff of the NRC. The Kemeny Report was quite specific about these problems:

> "...we have seen evidence that some of the old promotional philosophy still influences the regulatory practices of the NRC" (p19)
> "...the evidence suggests that the NRC has sometimes erred on the side of the industry's convenience rather than carrying out its primary mission of assuring safety" (p19)
> "There seems to be a persistent assumption that plants can be made sufficiently safe to be 'people-proof'" (p20)
> "We do not see evidence of effective managerial guidance from the top, and we do see evidence of some of the old AEC promotional philosophy in key officers below the top" (p20).

The Kemeny Report was hopeful that the reorganization of the NRC would begin a change in attitudes from the top down. A coherent plan for dealing with these difficult behavioral problems has not been forthcoming, yet obviously substantial changes are critical to a strengthened regulatory performance. The behavior of the NRC since the accident suggests, not surprisingly, that the pre-accident attitudes are proving difficult to extirpate. A scant five months after the accident and on the eve of the Kemeny Report, the NRC staff advised that technical fixes had so reduced the likelihood of a repeat of the TMI accident that new operating licenses could be issued even though the design basis of new reactors might be inadequate to control the potential consequences of the estimated amount of hydrogen released into containment at Three Mile Island. This action led the new Nuclear Safety Oversight Committee to observe in a letter to the President that a "business-as-usual" attitude continued to exist at the NRC. Also of concern was an NRC licensing reform plan announced in March 1981, that would restrict the role of intervenors and limit their access to information.

The response of the NRC, in summary, while probably improving its regulatory performance in a number of limited areas, has failed to resolve the need for more effective top leadership or the ingrained attitudes inimical to safety in the professional staff. It continues to be preoccupied with formal, specific pro-nuclear regulations to individual problems, leading one Kemeny Commissioner to conclude that "the NRC shows little recognition of the fundamental flaws in its approach to reactor safety"(Pigford 1981, p48).

The Kemeny Report also recommended that a new independent committee be instituted whose purpose would be "to examine, on a continuing basis, the performance of the [NRC] and of the nuclear industry in addressing and resolving important public safety issues associated with the construction and operation of nuclear power plants, and in exploring the overall risks of nuclear power" (p62). President Carter established the Nuclear Safety Oversight Committee in March 1980, under the chairmanship of Governor Bruce Babbitt of Arizona, and in its first year of existence the committee concentrated on post-TMI responses, issuing three letter reports on the NRC's Action Plan, iodine release in nuclear accidents, and emergency planning and response. The committee has to date played a useful oversight function but its future in the Reagan administration is in doubt.

Of equal or greater significance to the public institutions is the impact of the Kemeny Report on industry itself, especially the utilities that manage the operation of nuclear power plants. The report found far-reaching problems in the role of the utilities, warning that "the nuclear industry must dramatically change its attitudes toward safety and regulations", and that it must also "set and police its own standards of excellence to ensure the effective management and safe operation of nuclear power plants" (p68).

In fact, the major elements of industry response were set in motion well before even the appointment of the Kemeny Commission. The structure of this response in shown in Figure 2. Within two weeks of the accident, the four major industry groups—the American Public Power Association, the Atomic Industrial Forum, the Edison Electric Institute, and the National Rural Electric Cooperative Association—joined to create a policy task force (the TMI *ad hoc* Nuclear Oversight Committee) to address the safety issues raised by the accident. The seven subcommittees formed to develop policy recommendations indicated by the concerns immediately identified were:

emergency response planning;
operations;
systems and equipment;
post-accident recovery;
safety analysis considerations;
control room design;
unresolved generic safety issues.

Most committees reported their findings in September 1979; these findings also formed the industry's contribution to the NRC's Lessons Learned Task Force.

In a statement issued some three months after the accident, Floyd Lewis, chairman of the Industry *ad hoc* Committee, could point to three new institutions already begun or planned, as well as a wide range of other utility responses (US Congress, Senate Committee on Interior and Insular Affairs 1979). Approximately one month after the accident, the Nuclear Safety Analysis Center was established to conduct technical analyses of the accident, to interpret the lessons to be learned, to

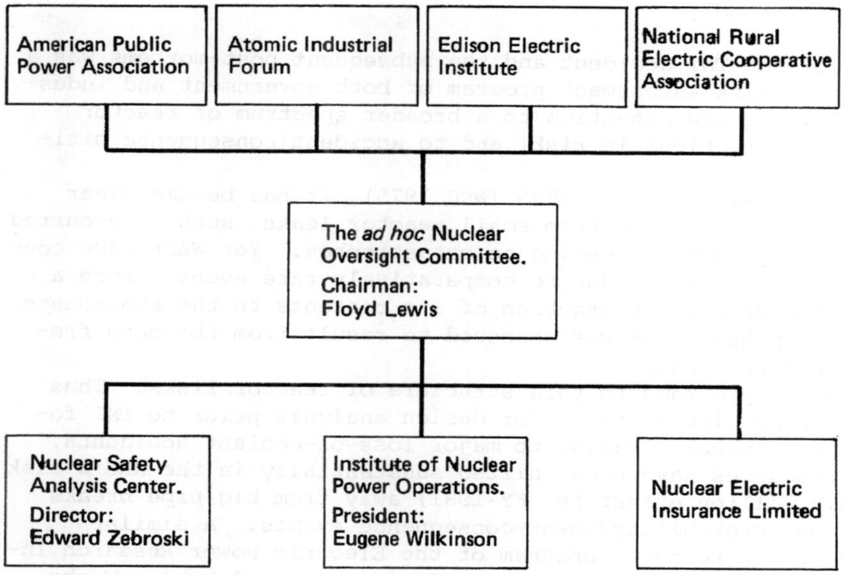

Figure 2. Response effort to the TMI accident by the nuclear industry.

develop strategies to prevent such accidents in the future, and to address generic safety issues. Financed by the utilities, the center has a professional staff of 50, and has so far completed a detailed technical analysis of the TMI accident, developed a priority system for needed safety changes, initiated a program on the testing of relief valves, conducted studies of a computerized database of 22 000 reactor operating failures (licensee event reports), instituted a computerized communications system linking 60 utilities for rapid dissemination of and requests for information, and conducted case studies of specific safety problems (e.g., the loss of electrical power to non-nuclear instrumentation at the Crystal River nuclear plant).

To deal with the "people problems" apparent in the TMI accident, in June 1979 the industry announced its intention to establish the Institute of Nuclear Power Operations. Chauncey Starr, with an advisory group drawn from the Navy's nuclear power program, the National Aeronautics and Space Administration, and airline safety, developed the mandate and structure of the institute. With a budget of $11 million in 1980, a projected staff of 200, and participation by 55 utilities, the institute will develop benchmarks for excellence in nuclear operations, conduct evaluations of individual utilities (some six completed by January 1981) with the goal of an annual evaluation for each plant, formulate educational and training requirements for operating personnel, and accredit training organizations. The utilities have also cooperated to create a mutual insurance plan to apply to the extraordinary costs accruing to a utility experiencing a major nuclear accident.

Taken together, these new institutions appear to represent a significant upgrading in industry's capacity and effort to manage nuclear power plant safety. It is noteworthy that the institutions emerged from industry's response to the accident and preceded the recommendations of the Kemeny Report. As with the NRC, the extent to which attitudes have changed among the rank-and-file of industry professionals is unclear. Certainly, as the response to TMI becomes institutionalized there is a temptation to believe that the outstanding problems are behind us and to declare nuclear power safe rather than to work to make it safer.

6. RISK ASSESSMENT AND MANAGEMENT

An important impact of the TMI accident and the subsequent post-mortems was a reorientation in the overall risk assessment program of both government and industry. Three key changes involve new attention to a broader spectrum of reactor accidents, to man-machine interactions in risk, and to accident consequence mitigation.

Since the publication of WASH 1400 in 1975 (NRC 1975), it has become clear that most postulated accidents will come from small reactor leaks, such as occurred at TMI, and not from large pipe breaks leading to catastrophes. Yet WASH 1400 concluded that the greatest public risk is due to comparatively rare events where a postulated melted core releases a large fraction of its contents to the atmosphere. Only a minor portion of the public risk was assessed to result from the more frequent melts leading to small releases.

The NRC has oriented its programs to this structure of reactor risks. Thus its risk assessment program and its criteria for design analysis prior to TMI focused heavily upon large pipe breaks leading to major loss-of-coolant accidents. Since the TMI accident, resources have been shifted substantially in the NRC's risk research program (a $231.9 million effort in FY-1981) away from big pipe breaks and transients toward higher-probability/lower-consequence events. A similar change has also occurred in the research program of the Electric Power Research Institute Nuclear Power Division (Senate Committee on Environment and Public Works 1980, p338; see Bibliography).

A second major change has been the allocation of significant new attention to human error as an ingredient in reactor accidents. WASH 1400 was quite inadequate in its attention to this issue, largely assuming human error rates taken from industrial situations assumed to be similar to nuclear power plants. In fact, a number of analyses, several of which were available before TMI, have demonstrated the importance of human error in reactor risks:

- (i) NRC official Merrill Taylor informed the Lewis Commission in 1978 that 50-85 percent of the hypothetical safety system failures he had examined in detail would be caused by humans (Sugarman 1979, p62).
- (ii) The 1978 German reactor risk study found that human failure was responsible for two-thirds of all risks.
- (iii) About 20-50 percent of all licensee event reports are due to human error (Sugarman 1979, p63).
- (iv) The NRC reports that in about 1 percent of all licensee event reports (about 25 incidents per year) there are indications that a safety feature has been seriously compromised or made unavailable by human error (Sugarman 1979, p63).

Despite this evident need to conceptualize reactor operations as a man-machine system, the NRC's approach to reactor risks remained strongly equipment-centered. Also, the NRC preoccupation with large-scale accidents ensured a neglect of human factors since such accidents require extremely fast reaction, thereby accentuating the role of automatic control through equipment.

It is not surprising, therefore, that a detailed post-accident review of the NRC's regulations, regulatory guides, and standard review plan found "no examples of criteria written with a clear intent to include human engineering considerations in the licensing and regulatory system" (NRC Special Inquiry Group 1980, Vol. II, Part II, p345).

The Kemeny Commission, along with the other accident post-mortems, was quite direct as to the significance of human behavior:

> "...as the evidence accumulated, it became clear that the fundamental problems are people-related problems and not equipment problems." (p8)
> "...wherever we looked, we found problems with the human beings who operate the plant, with the management that runs the key organization, and with the agency that is charged with assuring the safety of nuclear power plants." (p8)

"The most serious 'mindset' is the preoccupation of everyone with the safety of equipment, resulting in the down-playing of the importance of the human element in the nuclear power generation. We are tempted to say that...what the NRC and industry have failed to recognize sufficiently is that the human beings who manage and operate the plants constitute an important safety system." (p10)

Since the accident and the assessments, a number of efforts have been made to take human factors into account in nuclear risk assessment and management. The NRC has established a new Division of Human Factors and has restructured its risk assessment program to give greater emphasis to human error. The new Institute for Nuclear Power Operations has undertaken a major effort to improve the training of reactor operators, the single issue that most worried the Kemeny Commission. Since the accident, all 2500 licensed reactor operators in the US have gone through the TMI accident sequence on training simulators. Industry's new Nuclear Safety Analysis Center is evaluating human response in making design and instrumentation changes in reactor control rooms. Again, most of these issues were apparent and the response began prior to the publication of the Kemeny Report.

Despite these encouraging changes, questions still remain as to the adequacy of response, particularly in the NRC. In 1980 testimony before the Nuclear Safety Committee, Saul Levine, the former director of NRC risk evaluation studies, complained that "...the Agency is still grappling with equipment problems. Equipment is the be-all and end-all", and that the NRC continues to give insufficient attention to research on human factors (*Inside NRC* 1 December 1980, p13). Also, the NRC has decided against taking a lead role in the training of operators, unlike, for example, the Federal Aviation Administration's example with the training of pilots. It is questionable whether human behavior will ever become internalized into nuclear safety regulation without the direct role of the commission in such issues and a staff capability to allow in-depth analyses. The performance of the commission in socio-economic issues is instructive in this regard where, despite five years of continuing criticism of its inadequate analyses, the agency has yet to develop the requisite capability.

The third change in risk assessment involves a new focus on accident mitigation. Traditionally, this has also been an area badly neglected both by industry and the NRC. In no small measure, this is due to the widespread assumption that a serious reactor accident simply would not happen, and to the possible expenses involved in retrofitting requirements.

The Kemeny Report dealt primarily with emergency response, and the changes resulting from its recommendations will be considered below. Suffice it to note here that accident mitigation options are present and may have considerable potential for overall risk reduction. Consider, for example:

(i) Better containment building designs with filtered release systems to prevent containment failure due to internal over-pressure could, according to Von Hippel (1979, p79) be installed at both new and existing reactors at less than 1 percent of their replacement costs.

(ii) Means are available for interdicting the flow of contaminated water in the event of a melt-down from beneath reactor containment buildings to nearby water bodies, but no appropriate preparations have been made by the NRC.

(iii) Underground or more remote siting could substantially reduce the consequences of a major accident.

(iv) Thyroid blocking by use of potassium iodide pills could reduce thyroid tumor cases by as much as 100 000 over an area extending 200 miles downwind from the release source (Von Hippel 1979, p79).

Belatedly, the NRC is giving new emphasis in its risk assessment program, although most attention has focused to date on emergency planning. However, the recent findings that substantially less volatile iodine may be released in a reactor accident than was previously assumed may detract from accident mitigation analyses.

7. SITING AND EMERGENCY PREPAREDNESS

The TMI accident elevated siting and emergency preparedness into primary areas of nuclear safety concern. In regard to siting, the Kemeny Report concluded that the entire concept of a "low-population zone" on which the NRC based its siting criteria was flawed, and recommended:

> "In order to provide an added contribution to safety, the agency (i.e., the NRC) should be required, to the maximum extent possible, to locate new power plants in areas remote from concentrations of population. Siting determinations should be based on technical assessments of various classes of accidents that can take place, including those involving releases of low doses of radiation". (p64)

The siting problems of US nuclear power plants are the result of the location of plants progressively closer to cities over time as a means of reducing electricity transmission costs. Engineered safety features were substituted for remote siting as the primary means of risk reduction. Catastrophic (so-called class 9) accidents were judged to be of such low probability as not to merit inclusion in the siting criteria. As a result, power plants came to be sited very close to major metropolitan areas.

Immediately following the TMI accident, the NRC appointed a special Siting Policy Task Force to reconsider siting regulations. The task force reported to the NRC commissioners in August 1979 (two months prior to the Kemeny Report), calling for an abandonment of the principle of siting based on projections of potential dose commitments. Instead, there should be standardized fixed boundaries for exclusion areas and low-population zones. The report also recommended divorcing reactor designs from siting, thereby re-establishing siting as a major "defense-in-depth" factor. Final action is yet to be taken on the proposals, but there is substantial industry resistance as well as opposition from Europe and Japan, where such remoteness is unachievable, although West Germany is currently assessing underground siting.

Unresolved in the siting proposals are the power plants currently operating or on order. The NRC evaluation of 104 existing nuclear power plant sites found that about 30 failed to meet the new siting criteria, with Indian Point (near New York City) and Zion (near Chicago) being considered particularly bad (*Nucleonics Week* 25 October 1979, p3). The new siting proposals recommended "grandfathering" existing sites and compensating by emphasis upon emergency planning and additional engineered safeguards (e.g., core ladles). In short, the changes will do little to improve the overall siting of nuclear power reactors for the next few decades (or for a 100-150 GWE system).

Turning to emergency planning, the TMI accident demonstrated quite conclusively that none of the responsible parties was prepared for a major nuclear accident. The utility was unprepared to deal with the radiological aspects of the accident, the response by the NRC was disorganized and confused, and the local governments in the power plant region had no emergency plans to provide adequately for evacuation. The neglect within the NRC is evident from the fact that only three full-time professionals and one secretary out of 2500 NRC employees worked on emergency preparedness issues prior to the TMI accident (*Nucleonics Week* 8 November 1979, p10). The Kemeny Commission found few grounds for optimism:

> "The response to the emergency was dominated by an atmosphere of almost total confusion. There was lack of communication at all levels. Many key recommendations were made by individuals who were not in possession of accurate information, and those who managed the accident were slow to realize the significance and implications of the events that had taken place". (p17)

Although the commission noted a number of problem areas, it made no clear recommendations. Much of the response and outpouring of documents on the accident and the various post-mortems, including the Kemeny Report, centered nonetheless upon improving emergency response. By the time the Kemeny Report appeared, the Atomic Industrial Forum had developed a model plan for emergency response by the nuclear industry, calling for four well coordinated but independent emergency centers near the site, interconnected with reliable lines of communication. The Institute for Nuclear Power Operations has also focused upon training reactor operators in effective emergency response. The NRC has instituted a series of changes to improve its capabilities in this area, the most important of which are as follows:

(i) Six teams have been established to assess the emergency planning and preparation of every operating nuclear power plant. In addition, the new resident inspector at each plant has well defined duties during an accident.

(ii) The chairman of the NRC has been provided with clear lead responsibility during emergencies and an emergency management coordinator has been established to work with him.

(iii) Guidelines have been provided to licensees and the states defining classes of emergencies and outlining appropriate actions.

(iv) The NRC's operations center in Bethesda has been upgraded, and dedicated reactor operations telephone lines exist to each facility, with extensions in each control room. The NRC has also prescribed a data link between the plant's control room and a technical support center located somewhere on site. Consideration is under way for the data link to extend also to the NRC's Bethesda center.

(v) The licensing of new plants has been made conditional on the development of acceptable emergency plans.

In short, substantial improvements have been made in both industry's and the NRC's emergency response capability, although much remains to be done.

More problematic is the role of the states and local governments. The Federal Emergency Management Agency has lead responsibility in this area and in June 1980, in accordance with a Presidential directive, completed a review of state emergency plans for operating nuclear reactors (US Federal Emergency Management Agency 1980). The agency has mandated that a state should be able to notify every person within a 10-mile radius of a nuclear power plant that a nuclear accident has occurred and indicate what actions the person must take for personal safety. State governors hold the authority, unless delegated to local government, for ordering evacuation. By early 1981 it was apparent that, despite evident progress, many problems remained: the ability of utility officials to notify relevant state and local authorities promptly, is in doubt; a 1981 survey by *Nucleonics Week* indicated that most states are not yet prepared for an emergency and have, in fact, taken "an uncertain series of steps in notification procedures"; and confusion remains as to who has authority to do what during an emergency (*Nucleonics Week* 29 January 1981, pp2-3).

One final unresolved issue two years after the accident is the availability of potassium iodide pills in areas surrounding nuclear plants. The Kemeny Commission recommended that:

> "An adequate supply of the radiation protective (thyroid blocking) agent, potassium iodide for human use should be available regionally for distribution to the general population and workers affected by a radiological emergency". (p75)

The NRC has, as yet, reached no decision on the issue. The Federal Emergency Management Agency's position is that some $100 000 worth of potassium iodide should be stockpiled in four locations, but that distributional decisions be left to the states. In addition, the NRC staff paper on the issue advocates the availability of potassium iodide for hospitals and prisons near power plant sites. The lack of a clear policy some two years after the accident is regrettable, particularly given the routine distribution within 10 miles of a reactor in the UK and the recent Swedish decision in favor of dissemination.

8. CONCLUSIONS

Although any definitive statements must await implementation of ongoing changes and responses to future crises, the authors reach four tentative conclusions about societal response to the TMI accident and the Kemeny Report.

(1) Media coverage of the Kemeny Report showed a better performance by the scientific media than newspapers (at least as indicated by the *New York Times*), with the latter suffering from imbalanced attention to certain proposals to the neglect of other equally (or more) important findings, inordinate attention to leaks and investigatory work prior to the issuance of the report, and a lack of sustained analysis of the report itself and subsequent societal responses.

(2) The response by industry has been most timely and effective; regulatory responses have been more delayed and uneven; and the mass media have failed to respond to recommendations for change. The overall regulatory response has also been heavily dependent upon the role of industry. The long-term effect of the accident may be to further self-regulation in nuclear power.

(3) The changes instituted by industry and government have tended to address obvious gaps and specific problems apparent in the accident. The more fundamental and integrative problems of capability and attitudes which formed the primary concern of the Kemeny Commission and the need for new initiatives and ideas remain essentially unaddressed in the TMI response. One Kemeny Commission pro-nuclear member recently noted that "industry's concern with meeting the formalism of NRC regulations is still inhibiting and throttling new ideas and technical innovations more directly related to safety" (Pigford 1981, p47).

(4) The continuing failure to resolve the basic institutional problems of the NRC in the context of an anti-regulatory Reagan administration points toward a continuing vacuum in societal leadership for nuclear safety in the United States.

ACKNOWLEDGMENTS

The authors are indebted to the supprt of the Beijer Institute of Stockholm and the Swedish Energy R and D Commission. All responsibility for conclusions and any errors resides, of course, solely with the authors.

REFERENCES

Bertin, L. (1980) Harrisburg revised: TMI one year later. Ascent 1:17-23.

Bradford, P. (1980) Reasonable assurance, regulation, and reality. Speech before the ALI-ABA Course of Study on Atomic Energy Licensing and Regulation, September.

Bacher, D.E., and Carlson, D.D. (1980) Changes in the NRC's emergency response capability since Three Mile Island. Chapter 1 in: US Congress, Senate Committee on Environment and Public Works, Subcommittee on Nuclear Regulations. Staff Studies. Nuclear Accident and Recovery at Three Mile Island: A Special Investigation. 96th Congress, 2nd Session. Washington.

Donnelly, W. '., and Kramer, D.S. (1980) Nuclear Power: Three Mile Island Accident—Congressional Response and Investigations. US Library of Congress, Congressional Research Service, Archived Issue Brief 1B79097. Washington: Library of Congress.

Kramer, D.S. (1980) Reactor operator training—the utilities' role. Chapter 4 in: Part II US Congress, Senate Committee on Environment and Public Works, Subcommittee on Nuclear Regulations. Staff Studies, Nuclear Accident and Recovery at Three Mile Island: A Special Investigation. Washington: USGPO pp469-537.

Pigford, T.H. (1981) The management of nuclear safety: A review of TMI after two years. Nuclear News 24:41-48.
Sugarman, R. (1979) Nuclear power and the public risk. IEEE Spectrum 16:59-79.
Von Hippel, F. (1979) Physicist suggests way to mitigate effects of nuclear accident. IEEE Spectrum 16:79.

BIBLIOGRAPHY

Critical Mass Energy Project (1980) TMI: One Year in Retropsect. Washington.
Edison Electric Institute (1980) One Year After Three Mile Island: A Report to the President and the American People. The Institute.
US Congress, House Committee on Interior and Insular Affairs, Subcommittee on Energy and the Environment (1980) Industry's Response to the Accident at Three Mile Island. Oversight Hearings. 96th Congress, 1st Session, September. Washington.
US Congress, House Committee on Interstate and Foreign Commerce, Subcommittee on Energy and Power (1980) NRC's Response to the Report of the President's Commission on Three Mile Island. Hearings. 96th Congress, 1st Session, September. Washington: USGPO.
US Congress, House Committee on Science and Technology, Subcommittee on Energy Research and Production (1980) Nuclear Power Plant Safety after Three Mile Island. Report. 96th Congress, 2nd Session. Washington, USGPO.
US Congress, House Committee on Science and Technology, Subcommittee on Energy Research and Production (1980) Plans for Improved Safety of Nuclear Power Plants Following the Three Mile Island Accident. 96th Congress, 1st Session, September. Washington: USGPO.
US Congress, Senate Committee on Environment and Public Works, Subcommittee on Nuclear Regulation (1980) Report of the President's Commission on the Three Mile Island Accident. Hearings. 96th Congress, 1st Session, October. Washington.
US Federal Emergency Management Authority (1980) Report to the President: State Radiological Emergency Planning and Preparedness in Support of Commercial Nuclear Power Plants. Washington: The Agency.
US General Accounting Office (1980) Do Nuclear Regulatory Commission Plans Adequately Address Regulatory Deficiencies Highlighted by the Three Mile Island Accident? Washington: General Accounting Office.
US General Accounting Office (1980) Three Mile Island: The Most Studied Nuclear Accident in History. Washington: General Accounting Office.
US Regulatory Commission (1979) NRC Views and Analysis of the Recommendations of the President's Commission on the Accident at Three Mile Island. NUREG-0632. Washington: NRC.
US Nuclear Regulatory Commission (1980) Clarification of TMI Action Plan Requirements. NUREG-0737. Washington: NRC.
US Nuclear Regulatory Commission (1980) Modification of the Policy and Regulatory Practice Governing the Siting of Nuclear Power Reactors. Washington: NRC.
US Nuclear Regulatory Commission (1980) Plan for Developing a Safety Goal. NUREG-0735. Washington: NRC.
US Nuclear Regulatory Commission, Special Inquiry Group (1980) Three Mile Island: A Report to the Commission and to the Public. NUREG/CR-1250 (2 volumes). Washington: NRC.
US President's Commission on the Accident at Three Mile Island (1979) The Need for Change: The Legacy of TMI. Washington: USGPO.

Decision Making in Environmental Crisis Situations

James E. Dooley
Institute for Environmental Studies and Faculty of Management Studies, University of Toronto, Toronto, Canada

1. INTRODUCTION

1.1. Crisis Situations, Decisions, and Institutions

This paper discusses decision making in situations where an uncommon event occurs that threatens the safety of the public or the environment and for which authority structures and decision procedures are not routinely available. In order to focus the paper, the discussion will be limited to "environmental crises". This term, to be defined later, will include such man-made or natural situations as fires, hurricanes, avalanches, earthquakes, burst dams, floods, hazardous material spills, epidemics, snowstorms, large-scale food poisoning, and explosions. Excluded will be social unrest, economic crisis, military attack, and personal or family crises. The word crisis as used in this paper is synonymous with disaster or emergency.

Decision making is part of the management of such crisis situations. These almost always involve risk so that crisis management can be treated as a special case of risk management. There are many decisions involved through the stages of preparedness, crisis, and a new preparedness. This paper restricts attention primarily to the decisions that have to be made during the crisis.

The paper was stimulated by an environmental crisis that occurred in Mississauga, Ontario, in November 1979 (Burton 1981, Burton *et al.* 1981, Whyte *et al.* 1980) when 24 cars of a train became derailed, 19 of which carried hazardous materials. Three cars carrying propane exploded, caught fire, and burned for three days. An adjacent car carrying 90 tons of chlorine ruptured. A community of 226 000 people lived within the area where lethal concentrations could occur if a cloud of chlorine gas should escape and diffuse. The primary problem was to deal with the risk to human life by various evacuation options. The second problem was to contain the fire and chlorine gas in some way.

The research was undertaken to provide a context for examining decision processes in crisis situations and their relationship to institutional aspects of risk management. The work involved developing a set of characteristics for the decision process. These were developed in part from an examination of the Mississauga case. The characteristics, which are described in Section 2, are believed to have utility for comparative study. Some of the decision problems and their organizational setting are discussed in Section 3. In the final section a number of issues are identified that deal with the relationship between decision processes, the institutional setting, and preparedness.

1.2. Crisis Research

Results from a symposium on international crises are reported in Hermann (1972 a,b). Although international crises are not considered in the present work, the results of that symposium make several applicable suggestions on crisis research. Hermann suggests that the lack of theory on crisis is due to three factors: case histories rather than comparative studies are presented, so that knowledge accumulation does not take place, the respective authors tend to believe that each crisis is unique, and the concept of crisis as a phenomenon is not fully understood. There is no general agreement on how crises should be defined and characterized. It was found that appropriate definitions depended on the perspective and kinds of issues to be explored. For instance, Hermann suggests two viewpoints on the definition: the systemic and the decision making. The systemic defines a system in terms of a number of variables, and hence "...a crisis is a situation that creates an abrupt or sudden change in one or more of the basic system variables". The decision making approach, using the policy makers' perceptions of the crisis as a basis for definition is applicable when the concern is with the decision process and authoritative structure. Hermann's decision making approach has three defining characteristics: a threat to valued goals, restricted decision time, and surprise. Milburn (1972) offers criteria with which to judge whether or not a situation is a crisis: "situations become crises when (1) they concern values identified by the threatened policy makers as significant; (2) they are unexpected so that there is no set of plans or any existing program to handle them; and (3) they involve a relatively short time in which to decide to act before loss to values will occur". Robinson (1972) suggests classifying crises by two definitions: (i) substantive crises, based on the content of a particular policy, problem or situation, and (ii) procedural crises, based on a general class of generic characteristics.

A structure of decision making (coping) behavior in crisis situations has been developed by Janis and Mann (1977), from a review of research on psychological stress. They suggest that the decision making unit asks four basic questions which leads to five types of coping behavior according to whether the answers are "no" or "maybe" or "yes". The four questions and responses are given below.

1. Are the risks serious if I don't take protective action?
   - No leads to unconflicted adherence to the *status quo*.
   - Maybe or yes leads to question 2.
2. Are the risks serious if I do take the most available protective action?
   - No leads to taking the protective action in an unconflicted way.
   - Maybe or yes leads to question 3.
3. Is it realistic to hope to find a better means of escape?
   - No leads to pessimistically giving up the search for a better solution and to defensive avoidance behavior.
   - Maybe or yes leads to question 4.
4. Is there sufficient time to make a careful search for and evaluation of information and advice?
   - No leads to hypervigilant behavior resulting in taking the quickest, seemingly safe route.
   - Maybe or yes leads to vigilant behavior resulting in search, appraisal and contingency planning.

A schematic diagram from Janis and Mann (1977) illustrating the questions, behavior and consequences is reproduced in Figure 1.

Decision making under stress can result in pathological behavior as discussed by Smart and Vertinsky (1977). Their classes of crisis-specific pathologies that affect components of the decision making process are: narrowing the cognitive process, information distortion, group pathologies (e.g., groupthink; Janis 1972), rigidities in programming, and lack of decision readiness. Milburn (1972) discusses decision making in the context of the management of crisis. Billings *et al.* (1980) suggest that the way that the crisis is perceived affects the decision making behavior. They argue that a better understanding of perception will lead to better decision procedures. The work of Billings *et al.* suggests that the Janis and Mann model should treat the responses to the questions as perceived responses. The awareness of the possible difference between the "real" and "perceived" situation may lead to appropriate caution.

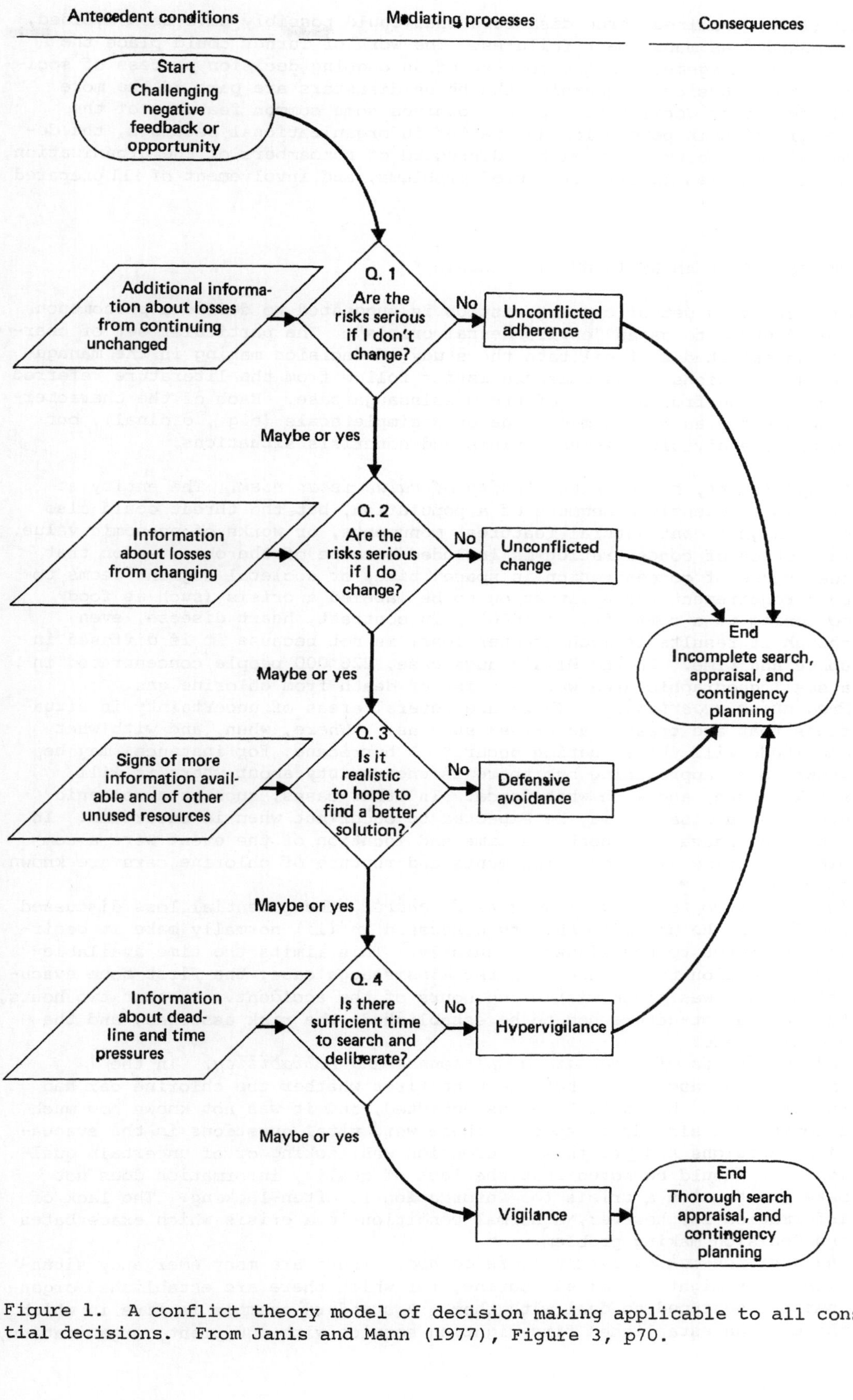

Figure 1. A conflict theory model of decision making applicable to all consequential decisions. From Janis and Mann (1977), Figure 3, p70.

Turner (1976) examined three disasters that could possibly have been avoided, and so are regarded as management failures. The work of Turner could place the problem of crisis management in the context of an ongoing decision process of society in which the decisions concerning the three disasters are part of the more general time series of decisions. Turner observes some common features of the failures: rigidities in perception and belief in organizational settings, the decoy problem (wrong problem dealt with), disregard of nonmembers of the organization, information difficulties in ill structured problems, and involvement of ill prepared people.

2. CHARACTERISTICS OF AN ENVIRONMENTAL CRISIS

In this section a set of characteristics is suggested to define a phenomenon that will be referred to as an "environmental crisis". The particular set of characteristics was selected to facilitate the study of decision making in the management of crisis situations. The characteristics follow from the literature referred to in Section 1.2 and from a study of the Mississauga case. Each of the characteristics can be treated as having magnitude on a simple scale (e.g., ordinal), but there is no precise division between crisis and noncrisis situations.

(i) *A significant, concentrated entity of value is at risk.* The entity at risk is frequently a segment of a population, but the threat could also be to significant natural features, monuments, or works of economic value. The notion of concentration is included because of the observation that the degree of concentration in space, time, or societal segment seems to be a requirement for a situation to be labeled a crisis (such as food poisoning to a community of 1000). In contrast, heart disease, even though it results in much greater loss, is not because it is diffused in space and time. In the Mississauga case, 226 000 people concentrated in a small geographic area were at risk of death from chlorine gas.

(ii) *Prediction uncertainty.* There are several areas of uncertainty in situations that are treated as crises such as: "Where, when, and with what magnitude will the situation occur?" A hurricane, for instance, may be known to be approaching but there is uncertainty about where it will strike, when, and with what force. In other cases, such as a volcanic eruption, a disaster may be expected to occur but when is uncertain. In the Mississauga case both the time and location of the event were a complete surprise although derailments and rupture of chlorine cars are known to occur.

(iii) *The lead time to decide (and act) is short.* The potential loss discussed in (i) and the unpredictability discussed in (ii) normally make it desirable to react to the situation quickly. This limits the time available for a decision to be made. In the Mississauga case, the first zone evacuation order was given within two hours of the accident. In that two hours, the command structure had to be established, the risk assessed, and the decision made.

(iv) *Information is of uncertain completeness and reliability.* In the Mississauga case it was not known at first whether the chlorine car had ruptured even though leakage was detected, and it was not known how much chlorine had already escaped. These were vital questions in the evacuation decisions and yet this information was lacking or of uncertain quality. It should be noted that the lack of quality information does not make a situation a crisis for information is often lacking. The lack of information is, however, a normal condition in a crisis which exacerbates the decision making problem.

(v) *The decision making authority is ad hoc.* There are many emergency situations that might be called routine, for which there are established organizations equipped to deal with them. A standard example is fire in which there is an established fire-fighting service with equipment, organization, and procedures in place. If, however, the fire should be of a magnitude or severity such that it exceeds the capability of the established service to deal with it routinely, then an *ad hoc* organization will result, which may be made up of components of the regular service. In many areas there are emergency-coping organizations, although these cannot be regarded as existing in a routine way for surprise disasters.

In the Mississauga case several organizations were established as part of emergency preparedness; for example, the Transport Emergency Assistance Plan was established in 1972 by the Canadian Chemical Producers Association. The Chlorine Institute, an association of chlorine manufacturers in North America, has a Chlorine Emergency Plan, which has a number of teams on call trained, in general, to deal with unplanned chlorine escapes. One such team was located at the Dow Chemical Plant, Sarnia, Ontario, some 270 km from the accident site. The regional government of Peel (Ontario is divided into a number of regions) has a Peel Regional Emergency Plan. It is interesting that in the Mississauga case, the local officials and politicians did not declare the situation an emergency, so the regional plan could not be invoked. Instead, the Police Disaster Plan was used and other agencies were called in on an *ad hoc* basis in general conformity with the needs and with some regard to the Peel Emergency Plan. Thus, in spite of the high level of preparedness in the area, in the final analysis the emergency was coped with by an *ad hoc* organization with little or no prior experience. The senior authority was the solicitor general for Ontario, Roy McMurtry, who assumed leadership and directed activities.

(vi) *Decision makers are not at risk.* In general the decision makers who have to respond to the crisis are not at direct personal risk from the cause. There may, of course, be individuals who escape the cause by evasive action, but in this characterization it is assumed that an organization not at risk is making the strategic decisions. In the Mississauga case a command post was set up near the accident site from which the decision makers operated. If chlorine had leaked, the command post would have had to be removed very rapidly, although this could probably have been done without loss of life. The decision makers are aware in such cases, however, that they may be at political risk in the post-crisis analysis and are likely to behave accordingly.

(vii) *The threat is nonhuman.* The characterization in this paper limits the crisis under consideration to those where the threat is from a nonhuman agent. The risk may be due to natural causes (fire, flood, etc.), or due to man-made works gone wrong (e.g., a chemical spill). This is the primary characteristic that separates environmental crises from personal crises or international incidents.

3. DECISIONS IN PERSPECTIVE

3.1. Crisis Stages

A crisis has three active stages: awareness, response, and return to a new normality. There are two further stages: prior preparedness, and revised preparedness. This view places crises in the context of a continuous process in which the three active stages become a particular event, albeit a spectacular one, that perturbs the ongoing process of survival. This view suggests that the crises must be studied in relation to the process of which crises form a part. The utility of this view is that a general theory of crisis that is stable over time is confounded by treating it relative to the state of the world at the time the crisis took place.

The awareness stage is concerned with the intelligence that would classify a situation as a crisis. The response involves establishment of the authority and the utilization of resources to contain the cause and to avoid its consequences. The third stage is concerned with dismantling the *ad hoc* authority and returning the resources and environment to what will now be regarded as a new normal state. Prior preparedness includes the warning systems and the established resources of people and materials to deal with the prior determined kinds of crises. Posterior preparedness is the new systems and resources that evolve from the experience of the last crisis. It is not claimed that the proper lessons are taken from the event; nevertheless, some interpretation will be made and a revised state of readiness instituted. As a case in point someone has observed that "we always prepare to fight the last war".

The process view leads to the obvious observation that in a crisis situation there are a large number of connected decisions. In order to study or prescribe, it is necessary to know which set of decisions is under consideration. To illustrate the range of decisions that can be involved, a number of decision problems are listed below. It will be observed that "selecting information" is classed as a decision, and also that recycling can take place among the decision problems.

Awareness stage

- Will there be an emergency?
- What level of crisis will be reached?
- What will be the risks?
- What entity (area, people, things) are at risk?
- Who should be advised (emergency organizations, management, public)?
- What authority should be established?

Response stage

- What information should be given to the public and how should this be communicated?
- Are local resources (materials, people) adequate and if not what further resources should be obtained and from which sources?
- What containment action should be taken?
- What evasive action should be taken?
- What resources should be allocated to containment and evasion?

Return stage

- Is it time for the *ad hoc* organization to turn over authority to the normal authorities?
- Can evasive provisions be terminated (e.g., evacuees return)?

3.2. Decision Processes

This section considers models and procedures applicable to the decisions needed during the crisis. Robinson (1972) has examined the literature from this viewpoint and has found that the decision literature rarely accommodates crisis. As a simple check on Robinson's critique the author examined the *International Abstracts in Operations Research* from 1977 to 1980. In the decision processes and decision theory sections, out of a total of 154 articles, none had the words crisis, emergency, disaster, or decision time in the title. The articles were not examined in detail, so there may well have been content on the subject. Nevertheless, the subject was not considered to be sufficiently important to be included in the titles. This tends to confirm Robinson's criticism.

Decision making in crisis situations appears to have been covered rather better in behavioral science (some literature was reviewed in Section 1.2; see also Janis 1974, George 1974). A considerable amount of work has also been done in decisions about personal crisis; see for example Coelho *et al.* (1974).

It may be that the analytical models arising from economics, operations research, or decision theory do not provide an adequate way to treat decision making in crisis situations. Any number of these models may be applicable, however, to decisions about preparedness to allocate resources, to locate facilities, and so on. What seems to have been neglected to a large extent is research on models applicable during crisis.

If the analytical model does not provide an adequate conceptual approach to crisis decision making, then is there an alternative approach that is better? The analytical approach has been found useful for well structured decision problems and can in many cases "make" the decision. Analytical models have been found very useful in providing insights into semi-structured decision problems through the process of model development and experiments with the model. The partial lack of structure naturally excludes some considerations from the model; if these are not to be ignored entirely then some other means has to be found to incorporate the other factors. One view is to construct a decision support system that provides for the flexible use of data and models that the decision maker then combines with judgment to decide. For a discussion of decision support systems see Dooley (1973), McCosh and Morton (1978), Keen and Morton (1978), and Alter (1980). The combination of judgment and models was expressed by Dooley (1973) in the form of a principle.

> "In the design of management information systems for decision support, the basic objective is to establish measures of effectiveness for the man-machine system and to seek an optimal sharing between humans and machines with respect to these measures."

As an illustration of the decision support system concept in crisis management, consider a nuclear power plant that has the potential to release a large amount of radioactive material. One of the critical things to know in the event of a release is the rate, direction, concentration, and diffusion of the radioactive material. A decision support system would have the means (developed ahead of time) to use information on the release and the prevailing weather conditions to predict the radiation threat. Judgment would then be used, in combination with the information on radiation, to decide on responsive action.

A procedural view of decision making, called procedural rationality, has been explored in a series of papers by Simon (1976, 1978a, b), in which he argues that the focus should be on examining the procedure by which a decision is reached rather than examining the substantively rational model (i.e., the "rational man" approach of economics). It may well be that an analytical model is the best procedural way to reach a decision but the focus is on the decision from a procedural viewpoint rather than from a model viewpoint. Simon (1976) gives a description of a decision making procedure that is procedurally rational:

> "There is now a large body of data describing human behavior in other problem situations of comparable complexity. All of the data point in the same direction, and provide essentially the same descriptions of the procedures men use to deal with situations where they are not able to compute an optimum. In all these situations, they use selective heuristics and means-end analysis to explore a small number of promising alternatives. They draw heavily upon past experience to detect the important features of the situation before them, features which are associated in memory with possibly relevant actions. They depend upon aspiration-like mechanisms to terminate the search when a satisfactory alternative has been found."

The approach described in this quotation could very well form the basis for studying decision making in crisis situations.

Simon (1978a) argues that in cases where the potential number of relevant considerations is great, only a few of the more salient ones are in the circle of awareness of the decision makers and that the set of considerations changes as learning takes place within the fluid organization. This condition is typical of decision making in crisis situations:

> "A general proposition that might be asserted about organizations is that the number of considerations that are potentially relevant to the effectiveness of an organization design is so large that only a few of

the more salient of these lie within the circle of awareness at any given time, that the membership of this subset changes continually as new situations (produced by external or internal events) arise, and that "learning" in the form of reaction to perceived consequences is the dominant way in which rationality exhibits itself."

"In a world where these kinds of adjustments are prominent, a theory of rational behavior must be quite as much concerned with the characteristics of the rational actors—the means they use to cope with uncertainty and cognitive complexity—as with the characteristics of the objective environment in which they make their decisions. In such a world, we must give an account not only of substantive rationality—the extent to which appropriate courses of action are chosen—but also procedural rationality—the effectiveness, in light of human cognitive powers and limitations, of the procedures used to choose actions. As economics moves out toward situations of increasing cognitive complexity, it becomes increasingly concerned with the ability of actors to cope with the complexity, and hence with the procedural aspects of rationality."

An interpretation of operations research/management science, artificial intelligence, and computational complexity as a form of procedural rationality is given in Simon (1978b). Simon refers to the three approaches as normative theories of procedural rationality which reflect the strong attachment that these approaches have to optimizing (or satisficing) with respect to stated criteria. He goes on to cite the parallel development in cognitive psychology which he refers to as a positive theory of procedural rationality. The limitations on human information processing are said to stem from low serial processing rates and the limitations of short-term memory. Simon argues that a positive aspect is the human ability to store large amounts of data in long-term memory and to recall this in response to stimuli.

The similarities between decision making in crisis situations and procedural rationality are immediate. For instance, the organization is fluid and has time to consider only a limited number of considerations (Hermann 1972b). The decision makers will make judgments based on past experience in similar situations. Models, if used at all during a crisis, will be justified only if they fit within the procedure adopted by the decision maker and can be applied within the time and information constraints.

3.3. Institutional Context of Crisis Decision Making

So far only the nature of decision problems in crisis situations have been discussed. In this section a brief discussion of a typical institutional context is given. Clearly, the organization will be formed to meet the particular crisis and will be affected by the existing relevant culture and emergency planning institutions. The hypothetical organization in this section will illustrate the institutional setting for the decision process.

In order to be effective the organization must consider both the decision and how it is to be implemented. In fact, the potential for implementation can often be an important consideration in choosing an alternative. Figure 2 shows a typical organization in which a committee or person (or both) assume command to direct the crisis response. The command group makes the major decisions on responding actions. An advisory group supplies information consisting of an appraisal of the situation, resources available, and so on. The command has the responsibility to arrange for resources to be made available to the extent needed or possible. The next part of the organization is the coordination unit which directs operations toward fulfilling the strategic decisions. Next there are the operations units, typically autonomous, which may have their own organization, material, personnel, and procedures such as the police, fire department, and ambulance service. Also, responding units

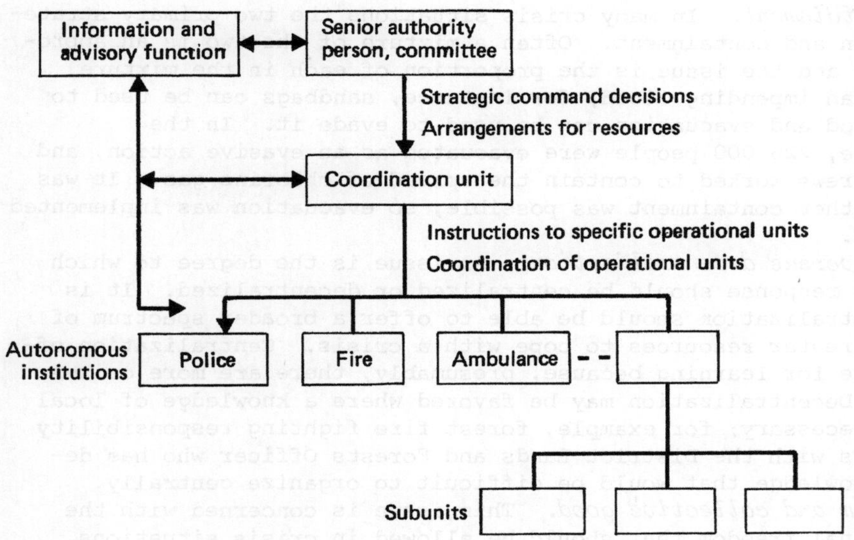

Figure 2. Typical crisis management organization.

may normally exist for the kinds of emergencies experienced locally. An example is forest fire fighting which, in Ontario, is routine for most fires. Special units may be formed to meet the needs of the particular crisis. A coordination problem is to get the autonomous units to function together effectively.

4. CRISIS DECISIONS, RISK, AND INSTITUTIONS

Mankind has been coping with environmental crises (disasters, emergencies) for centuries and has developed local coping behavior of varying quality for the classes of crises experienced. It is well beyond the scope of this paper to attempt even a cursory examination of the various institutions and decision procedures that they use. In raising a number of issues concerned with crisis (risk) management it is understood that these have been dealt with very effectively in some parts of the world and for some classes of crises. The issues are offered as topics for research. Existing cases of crisis response and institutions can be examined for generalizations about crisis management. This paper has focused on decision making during crises. In the issues that follow it will be apparent that the scope for decision making during crisis is conditioned by organization, resource, and decision preparedness. Consequently, most of the issues concern preparation as well as response during the crisis.

(a) *Prediction and response.* The issue is: "How should limited resources be allocated to prediction and to response for a particular class of crisis?" One extreme is to have extensive warning systems but no response capability; the other is to be surprised at the occurrence of each crisis but to have extensive resources to cope with them when they do occur. In Ontario for instance, a local type of emergency is forest fire. Prediction consists of surveillance of weather and forest conditions to assess the capacity for fires, and reconnaissance aircraft to spot them. The amount of reconnaissance increases as the conditions for fire become more critical. Response consists of the mobilization of forces to fight the fire; these can be smaller if the warning is given early enough. The mix of prediction and response systems has a significant bearing on preparedness and on the options open to decision makers when a crisis occurs.

(b) *Evasion and containment.* In many crisis situations the two primary strategies are evasion and containment. Often a mixture of the two is an appropriate strategy and the issue is the proportion of each in the mixture. In the case of an impending flood, for instance, sandbags can be used to contain the flood and evacuation can be used to evade it. In the Mississauga case, 226 000 people were evacuated as an evasive action, and fire fighting crews worked to contain the spread of chlorine gas. It was not certain whether containment was possible, so evacuation was implemented as a precaution.

(c) *Centralization versus decentralization.* An issue is the degree to which preparation and response should be centralized or decentralized. It is argued that centralization should be able to offer a broader spectrum of knowledge and greater resources to cope with a crisis. Centralization offers more chance for learning because, presumably, there are more crises to deal with. Decentralization may be favored where a knowledge of local conditions is necessary; for example, forest fire fighting responsibility in Ontario rests with the District Lands and Forests Officer who has detailed local knowledge that would be difficult to organize centrally.

(d) *Personal freedom and collective good.* This issue is concerned with the amount of personal freedom that should be allowed in crisis situations where such freedom may reduce the collective good. There are several examples of this; for instance, should a person be allowed to refuse an innoculation during an epidemic where refusal may result in that person infecting others? Can an individual refuse evacuation when conditions may put others at risk trying to rescue that individual? The issue could also be stated as individual freedom to accept a risk when in so doing greater risk is imposed on others. A related issue is how much individual risk should be accepted for the common good?

(e) *Ad hoc authority.* For "normal emergencies" an organization is developed over time. Through legislation and performance the organization commands authority and respect to deal with certain situations. When the situation exceeds the normal, an *ad hoc* organization is formed that may not have any prior experience of that type of emergency. The issue is: how can such an organization derive the authority and respect to command and coordinate the numerous groups involved in coping with the crisis? The problem becomes more acute when normally autonomous, experienced organizations (e.g. fire, police) are asked to serve under the *ad hoc* authority.

(f) *Decision making experience.* It has been argued throughout this paper that the analytical (substantively rational) approach to decision making is not adequate for decisions to be made under crisis conditions. It has been argued that procedures that combine judgment and experience with analysis are the appropriate model. Regardless of which view is favored, judgment and experience will form a large part of the decision process out of necessity. Experience will often be available within the operating units, but is not likely to be available in the command authority. The issue is to develop some form of decision preparedness. One observation is that people in political authority have appropriate experience in decision making in highly political environments; in marshaling resources, and in coordinating diverse groups. It has been argued that the political decision environment has a lot of characteristics in common with that of crisis situations, and the Mississauga experience tends to support this view. In Ontario there is a "lead" minister designated to head the response organization for each class of crisis (Timmerman 1980).

(g) *Return to normality.* Just as procedures, and possibly organizations, must be established to cope with a crisis, the reverse is to disband and return to normal. The problem is to decide when the crisis is over, what the new normal condition is, and how to return to it.

(h) *Risk management responsibility for crisis avoidance.* The nuclear industry can be used to illustrate this issue. The two extreme cases are: full government responsibility, and full industry responsibility. In the first case government would perform inspections and set regulations and procedures effecting risk in all phases of construction and operation, so that the industry's responsibility ends at compliance with the regulations and procedures. The other extreme is to set risk levels and leave it to the industry to decide how to achieve them. If the industry fails then it may be too late to do anything to avoid a disaster; they have failed but that does not help those who suffer. The issue is to decide on a balance between these two extremes.

(i) *Communication with those at risk*. Studies have shown, as reported in Janis and Mann (1977), that the public reaction to a slow build-up of a disaster is different from those to crises that have a more rapid onset. The public tends to treat the slow build-up as indicating lesser risk or less need to act promptly when advised. This is only one observation on behavior. The issue is: how much information should be given, when should it be given, and how should it be given? This is an important issue for those cases where some form of response by those at risk is needed. Related to this issue is the question of the public's right to know.

(j) *Information and communication system*. It is fundamental to effective decision making in crisis situations that information and communication systems are adequate to provide sufficient data as rapidly as possible. This obvious truism is particularly important in crisis situations where it may not be possible to feel one's way incrementally toward a solution (incrementalism). The problem for preparedness is to prepare by having information and communications systems in place for the surprise situations. Traditionally the established media have played a large role. More recently, the citizens' band radios have been important in rapid communication. The early warning systems of the military are further examples.

(k) *Resources*. Television programs and movies about crisis situations are instructive. It is amazing that endless numbers of skilled people, adequate equipment, and leaders are always available to cope with crises. What these dramatizations explore is how well we can cope when the resources are available and we work together. In fact, they use the absence or failure of a resource to build tension which further illustrates the importance of resources and their coordinated utilization. And yet, in reality, you cannot get a taxi when it is raining in New York. This paper does not attempt to provide a catalog of resources to cope with crisis situations. The issue is: what is the overall availability of resources (information, communication systems, material, human skills, and organization) to cope effectively with crises?

(l) *Crisis management preparedness*. Management of crises can be treated as consisting of an organizational structure, decision procedures, and resources. The issue is to prepare so that these can be brought together in an effective way to cope with crisis situations. Under crisis conditions it may be too late to start thinking about management so that preparedness requires that some thought be given to these matters ahead of time. One way is to have crisis exercises in which participants presuppose a given kind of crisis and simulate ways of coping. In Ontario, for instance, regular exercises are held to test the system for responding to nuclear emission events.

ACKNOWLEDGMENTS

The author wishes to thank Professor Ian Burton, Institute for Environmental Studies, University of Toronto, for suggesting the topic, for supplying information on the Mississauga case, and for reading the manuscript. The help of C.L. Kirby, Emergency Planning Coordinator, Office of the Solicitor General, Ontario, and L. Balogh, Flood Emergency Coordinator, Department of Lands and Forests, Ontario, is acknowledged with thanks. Appreciation is given to the National Research Council, Canada for providing findancial assistance for this research.

REFERENCES

Alter, S.L. (1980) Decision Support Systems: Current Practices and Continuing Challenges. Reading, Mass.: Addison-Wesley.

Billings, R.S., Milburn, T.W., and Schaalman, M.L. (1980) A model of crisis perception: a theoretical and empirical analysis. Administrative Science Quarterly 25:300-316.

Burton, I. (1981) What Happened at Mississauga? Paper presented to the WHO Regional Conference on Contingency Planning for, and Response to, Emergencies and Accidents Involving Potentially Toxic Chemicals, Bilthoven, Netherlands, 1981.

Burton, I. *et al.* (1981) The Mississauga Evacuation Final Report. Toronto: Toronto University, Institute for Environmental Studies.

Coelho, G.V., Hamburg, D.A., and Adams, J.E. (Eds.)(1974) Coping and Adaptation. New York: Basic Books.

Dooley, J.E. (1973) Management Decision Support Systems. Working Paper 73-11. Toronto: Toronto University, Faculty of Management Studies.

George, A.L. (1974) Adaptation to stress in political decision making: The individual, small group, and organizational context, in Coelho *et al.* (1974) pp176-245.

Hay, J., and Williams, S. (1966) A comparison of decision making performances in established and *ad hoc* groups. Journal of Personality and Social Psychology 3:214-222.

Hedburg, B.T., Nystrom, P.C., and Starbuck, W.H. (1976) Camping on seesaws: prescriptions for a self-designing organization. Administrative Science Quarterly 21:41-65.

Hermann, C.F. (Ed.)(1972a) International Crises: Insights from Behavioural Research. New York: Free Press.

Hermann, C.F. (1972b) Threat, time, and surprise: a simulation of international crises, in Hermann, C.F. (1972a) pp187-211.

Janis, I.L. (1972) Victims of Groupthink. Boston: Houghton-Mifflin.

Janis, I.L. (1974) Vigilance and decision making in personal crisis, in Coelho *et al.* (1974) pp139-175.

Janis, I.L., and Mann, L. (1977) Decision Making. New York: Free Press.

Keen, P.G., and Morton, S.S. (1978) Decision Support Systems: An Organizational Perspective. Reading, Mass.: Addison-Wesley.

McCosh, A.M., and Morton, S.S. (1978) Management Decision Support Systems. London: MacMillan.

Milburn, T.W. (1972) The management of crisis, in Hermann, C.F. (1972a).

Robinson, J.A. (1972) Crisis: an appraisal of concepts and theories, in Hermann, C.F. (1972a) pp20-35.

Simon, H.A. (1976) From substantive to procedural rationality, in S.J. Ratis (1976) Methods and Appraisal in Economics. Cambridge: Cambridge University Press.

Simon, H.A. (1978a) Rationality as process and as product of thought. Journal of the American Economic Association 68(2):1-16.

Simon, H.A. (1978b) On how to decide what to do. The Bell Journal of Economics 9(2):494-507.

Slovic, P., Fischhoff, B., and Lictenstein, S. (1977) Behavioral Decision Theory. Annual Review of Psychology 28:1-39.

Smart, C., and Vertinsky, I. (1977) Designs for crisis decision units. Administrative Science Quarterly 22:640-657.

Timmermann, P. (1980) Emergency Planning in Ontario: A Critical History and Analysis. Working Paper ERR-9. Toronto: Toronto University, Emergency and Risk Research, Institute for Environmental Studies.

Turner, B.A. (1976) The organizational and interorganizational development of disasters. Administrative Science Quarterly 21:378-389.

Whyte, A.V.T., Liverman, D.M., and Wilson, J.P. (1980) Preliminary Report on Survey of Households Evacuated During the Mississauga Chlorine Gas Emergency, Nov. 10-16, 1979. Emergency and Risk Research Working Paper 7. Toronto: Toronto University, Institute for Environmental Studies.

Main Gas Pipeline Route Selection Problems, Taking into Consideration Risk and Uncertainty Factors

Y.S. Oseredko, O.I. Larichev, and A.I. Mechitov
All-Union Designing Institute for the Gas Industry, Kiev, USSR

1. INTRODUCTION

At present, appreciable structural variations in the world energy situation are taking place. Petroleum, formerly considered to be the most valuable chemical-organic raw material and source of energy, is to a large extent being replaced by other energy sources such as gas, nuclear power, hydroelectric power, and coal. Natural gas is probably the next most important source after petroleum because of the availability of substantial world resources, ease of transportation, good technical-economic utilization characteristics, etc. Natural gas is also a valuable raw material for the chemical industry, and is widely used for obtaining important basic materials. These factors predetermine the rapid development of world gas output and trade. The USSR, according to the 1981-1985 five-year plan, is intending to increase its output of gas by 38-47 percent, to 600-400 billion m^3.

The harnessing of new gas resources and the construction of complexes for transportation, storage, and reprocessing of gas require large investments of capital and long time commitments. When considering such large-scale projects it is necessary to take into consideration economic, political, social, geological, and other factors involved, as well as questions of population safety, environment protection, reliability of transportation systems, etc. Thus the creation of a large gas complex is a complicated task, where an essential role is played by risk and uncertainty factors.

2. CHARACTERISTICS OF A GAS EXPORT COMPLEX

Among European countries the USSR is the only one which exports natural gas in significant quantities. However, the main gas producing regions in the USSR are in the northern, sparsely developed areas of the country, and are therefore considerable distances from the main industrial centers and potential foreign consumers. This has necessitated the construction of very long pipelines to transport the gas to consumers.

At present in the USSR liquefied natural gas (LNG) processing complexes, incorporating gas liquefaction plants, storage, and shipping facilities, are under preliminary development in the north of the European part of the USSR and on the Far Eastern coast. Both areas, however, are considerable distances from the largest gas deposits. For this reason a component part of any such complex must inevitably be the main gas transportation system. Here gas pipelines, which may be several thousand kilometers long, are the main factor determining the cost of the whole complex and its effectiveness. To a great extent this is because the pipelines have to be constructed through climatically severe, unpopulated regions, and analysis shows that the capital outlay required for such a venture may amount to 75-80 percent of the total cost of the complex.

The building of a gas pipeline is therefore a significant and often decisive element in a gas transportation complex designed to provide large-scale gas supplies. Because of this it is expedient to point out the numerous factors connected with the rational selection of gas pipeline routes, such as environmental conditions, socio-economic factors, the local population, and agreements between land-owners and administrative organizations.

When constructing a gas pipeline a guarding zone of 250-350 m from its axis is required to regulate the minimum distance from the pipeline to residential buildings, highways, farms, and other installations; with a route length of about 1000 km this amounts to 50-70 thousand hectares. Thus the main pipeline route will constitute a site of temendous size, the selection of which represents a very serious problem. The task being considered is also important practically. In the USSR the construction of 49.5 thousand km of gas pipelines for 1981-1985 is being scheduled. Fast development of pipeline transport, particularly for gases, is also characteristic of world economics as a whole. Thus, route selection problems for similar installations will become much more pronounced in the course of time (Belousov 1978).

3. FACTORS CONSIDERED IN PIPELINE ROUTE SELECTION

A gas pipeline several thousand kilometers long is a complicated and expensive technical system. Route selection depends upon natural climatic and economic-geographic conditions including the presence of topographic, geological, hydrological, natural, and artificial obstacles. We can classify the variety of natural features of the terrain along the route under the following headings: plains, deserts, swamps, permafrost, natural water barriers, and mountains. However, these headings do not include the whole range of factors which could affect construction and installation work, such as large tracts of forest, high water table, rocky soils, impassable swamps, formation of crevasses and thermokarsts in permafrost soils, deep and also planned recutting of river beds, landslides, etc. The list of headings can also be further subdivided into a series of categories that permit us to consider specific features of the terrain.

It is also necessary to take into account the quantity and size of populated areas, the effect on the route length of bypass lines, the increase in the amount of metal required to increase the pipeline wall thickness in order to safeguard the local population, and the rise in construction costs due to any demolition of dwellings and other buildings that may be necessary. It is also necessary to consider the prospective development of populated areas and cities in the next 25 years.

The quantity and quality of agricultural land have increased, so funds must be made available to cover the expenses of recultivation, reimbursement of production losses, restoration of irrigation systems, etc. Access to existing road and rail networks (to simplify the transporting of pipes, equipment, construction machinery, etc.), as well as access to existing pipelines along the route, communications, and power supply systems, all greatly reduce the amount of construction and installation work required for the building of a pipeline.

We can now specify the basic factors considered in the main gas pipeline route selection.

3.1. Presented Costs

This factor is the most common and universal estimation criterion, determined from the expression

$$C = K x + A$$

where K is the capital investment; x is the normative coefficient of capital investment efficiency (for industrial construction it is taken to be 0.12); and A is the annual maintenance cost. This basic criterion permits selection of a route from an initial to a final point which will require the minimum total capital outlay (equipment and labor) and maintenance expenses. However, it does not guarantee the selection of a truly optimal route because it can not take into account all environmental, social factors, etc., because these cannot be accurately estimated in advance (Goncharov and Oseredjko 1977).

3.2. Construction Times

This may be one of the most decisive factors in commissioning a gas pipeline. The duration of construction is partially dependent on the standards required, but can also be established by means of directive instructions. In general the preferred route alternative is that where appropriate construction organizations already exist and where seasonal transport routes are available, or where pipelines already exist, together with appropriate maintenance systems. Also, the factor of minimum change to existing construction technology, machinery, etc., is considered, as well as the availability of an adequate labor force.

3.3. Gas Pipeline Maintenance

In order to ensure reliable pipeline operation it is necessary to have access to all sections for preventive inspections, and repair work in the case of failure. Access is to some extent determined by environmental conditions in the pipeline area and by the development of the transportation system.

The reliability of maintenance mainly depends on natural climatic conditions along the route. In some cases, to ensure faultless operation of gas transportation systems in the most complicated and important sections (large areas of water or swamps, almost inaccessible mountain regions) laying two pipelines instead of one is standard practice, even though this means an increase in capital investment and can require gas supply reservation by means of underground storage in natural formations, etc.

3.4. Influence on the Environment

Construction of main gas pipelines, especially when they are three to four thousand kilometers long, has a great impact on the environment. Working on such a scale, partial forest clearance and agricultural production losses are inevitable, and are often evaluated without consideration of the long-term effects on the environment.

When laying a pipeline in highlands where there is a danger of landslides, this is not only a threat to the environment, but also affects pipeline reliability since landslides may be caused when slopes are cut to form terraces for the movement of construction machinery and for laying the pipeline. Disruptions to hydrological systems may occur; for instance, the construction of underground pipelines at insufficient depth may cause water channel deformation and drainage disruption that can adversely affect aquatic life and prevent accident-free running of the pipeline.

The environment is especially sensitive to human activities in the permafrost areas of the north. The main gas producing area of the USSR is in the north, so that the gas pipelines need to be constructed through frozen permafrost soils for considerable distances from the deposits to the LNG processing plant. To prevent disturbance of the permafrost and to ensure pipeline stability, gas cooling is used, wherein a specified temperature regime is maintained by an associated gas cooling station. However, deterioration of temperature control may occur after several years of operation, so that the influence of this on the fauna and flora of the area surrounding the pipeline must be evaluated when considering route alternatives.

Finally, factors associated with the installation process can have a major effect on the environment. For example, it is known that the noise of a gas turbine unit can have a very serious influence not only on the maintenance personnel and nearby inhabitants, but also on animals and their activities.

3.5. Connection with Regional Development Plans

The influence of gas pipeline construction on the population and economics of a region should also be taken into consideration. The building of large LNG complexes requires several thousand specialists and workers, including some from other countries, and the type of labor force required during the different construction phases may change sharply. The socio-economic influences on this influx on the populated areas close to the construction area may be important and should be carefully evaluated. When demolition of homes is necessary the problems of rehousing and selection of new residential areas arise. A subject of serious concern should be the provision of social and cultural facilities, services, etc., for the maintenance personnel and their families. The creation of such an infrastructure often involves considerable capital cost.

With balanced regard for all factors affecting the construction work and the existing socio-economic structure of the construction area, some negative consequences of construction may be compensated for by, for example, supplying gas to the area; introducing a centralized heating system for homes and agricultural installations using the waste hot water from compressor station gas turbine cooling systems; construction of new socio-cultural and life-enhancing facilities, communications systems, etc.

The influence of the gas pipeline on regional development plans often affects public opinion. The attitude of the local administrative bodies and the local population towards the approaching construction of the pipeline and the positive measures taken by these groups when making decisions on the allocation of land, permission for the various construction processes, etc., can greatly speed up the completion of the design and surveying work, as well as the construction process itself. Obviously this improves the economic viability of the project.

3.6. Construction Conditions

This factor is determined by geological, hydrological, topographic, and other conditions of the gas pipeline route selected, as well as the availability of existing infrastructure, construction bases, etc., in the region. It is necessary to consider it as an independent factor because it is important for the construction firm that also takes part in the process of route selection.

3.7. Population Safety

This is mainly ensured by keeping to the standard minimum distances from the main gas pipeline axis to populated areas, buildings, farms, highways, etc. (the guarding zone). However, this measure does not completely guarantee population safety in emergency situations. There are two ways of reducing the potential risks:

(i) by increasing the reliability of technological systems and installations;
(ii) by extending the gas pipeline guarding zone and distances from the various systems to the nearest populated areas.

It should be noted that regulations concerning violations of the guarding zone, and fire protection regulations in the USSR, are the most extensive in the world, but this affects some economic factors.

3.8. Special Permissions

Special permission requirements and limitations imposed by laws on the protection of nature, fish reserves, mineral resources, forestry, etc., and also regulations issued by sanitation and fire protection authorities, etc., are very important factors in pipeline route selection. Obviously when selecting a pipeline route it is necessary to take into consideration many factors, some of which can be expressed quantitatively, and others only in a qualitative, descriptive fashion. Initial data on the route alternatives may be available to varying degrees of accuracy. The construction of a pipeline involves vast expenditure and a design stage that may well continue for several years, yet the decision making cannot totally remove the element of risk. It is necessary to take into account the degree of reliability of gas pipeline maintenance, the conditions required for its safe functioning, its influence on the environment, etc. Thus, pipeline construction represents a good example of a multi-attribute task of decision making under conditions of risk and uncertainty.

4. AN EXAMPLE OF ROUTE SELECTION

The selection of an optimal route for the Kutaisi-Sukhumi gas pipeline, together with its branch pipelines to Poti and Batumi for gas supply to domestic and industrial consumers is considered here as a specific example. The basic factors that were taken into account, and the selection procedure, were as follows.

At the preliminary stage of the study (research, field inspections, preliminary agreements) three possible routes were selected: piedmont, median, and maritime (see Figure 1). In addition to the main gas pipeline, prospective branch pipeline routes to populated areas and other consumers were taken into consideration.

The piedmont version was the shortest route, passing through spurs of the Egriss ridge. The relief is heavily dissected with canyons carrying mountain rivers, and the elevation varies by about 700 m. Small villages are located in valleys along the route, so that demolition of homes would be inevitable. Otherwise it would be necessary to go through quite complicated mountain conditions to bypass them. Construction work would be fraught with the dangers of mud-laden torrents with karst land forms and landslides, and would be aggravated by the need to cut special "terraces" into steep slopes in order to move construction machinery and for laying the pipeline. The route is quite distant from populated zones, and the road system is not highly developed, so that pipeline surveillance and maintenance would only be possible with the help of helicopters.

The median version passes through populated zones. The relief is gentle with good geological conditions, and the road and rail systems are well developed. However, this version would require the greatest amount of building demolition and destruction of cultivated areas; it would also be necessary to build numerous crossings over artificial obstacles.

The maritime version was the shortest route, passing through the Kolkhida lowland area. The relief is fairly flat with considerable areas of alder woodland, and a well developed irrigation system. Possibly this route could have had to pass through swamps, making conditions along the route difficult, especially during rainy periods. This route would require very little building demolition or crop destruction but the situation would be quite different along the prospective branch pipelines. For the preliminary study the area to which the optimal route search was to be restricted was defined on topographic maps drawn up from aerial photographs.

5. COMPARISON OF THE THREE VERSIONS

After the three basic alternatives for the pipeline route had been considered it was possible to use correlation analysis using the criteria detailed above.

5.1. Presented Costs

The costs of construction depended upon the difficulties encountered in the laying of the gas pipeline (swamps and river crossings, terrace cutting on the mountain slopes, etc.) and its maintenance. Along with this, the costs arising from demolition of buildings (if it was impossible to go round them) were taken into consideration, as well as compensation payable if the route crossed cultivated land.

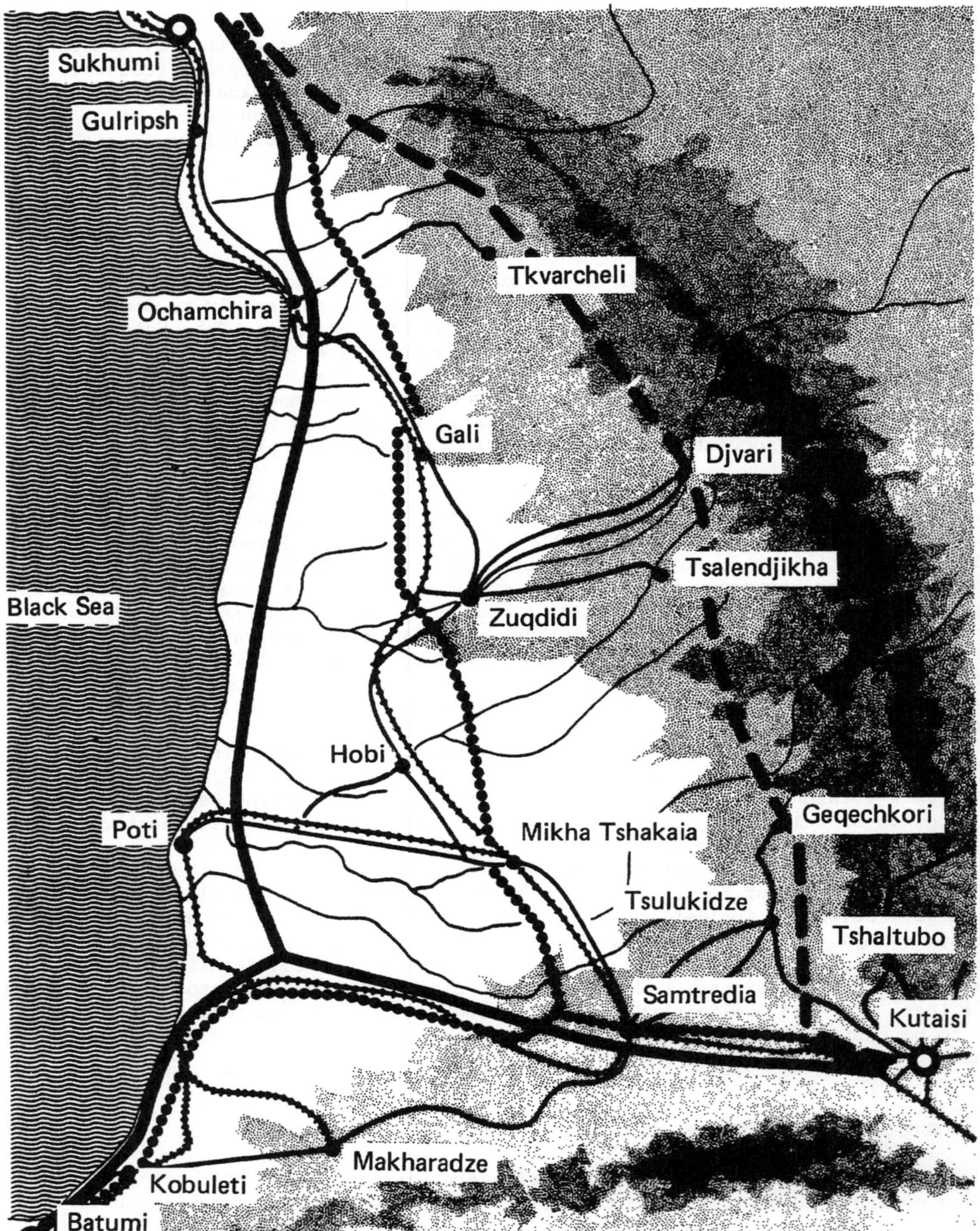

Figure 1. Scheme of the alternative routes for the Kutaisi-Sukhumi gas pipeline. Full line, maritime route; dotted line, median route; broken line, piedmont route.

These costs were determined by local administrators. According to the formula in Section 3.1 the respective costs of the three versions were as follows (in millions of roubles): maritime—8.9; median—9.5; and piedmont—10.8.

The pipeline construction process can be divided into two stages. The first stage is the construction of the main pipeline itself; the second is the laying of branch pipelines (from the main pipe to the consumers). These stages are not separated merely in time; they are also financed from different sources: the construction of branch pipelines is paid for by resources from the regional budgets. In view of this, two cost criteria must be considered separately: capital costs for the main pipeline, and those for the branch lines. Capital investment in the main route and the construction of the branch lines were, respectively (in millions of roubles): maritime route—31 and 9.5; median route—40 and 5; and piedmont route—46 and 5.

5.2. Construction Times

According to existing standards for gas pipeline construction the construction times for the different versions would not be expected to differ greatly. However, based on experience, the piedmont version would probably take much more time because of route laying difficulties. Along the maritime route construction delays might occur in marshy areas and when crossing three big rivers.

5.3. Gas Pipeline Maintenance

The most difficult route to maintain was recognized to be the piedmont version since access can only be achieved with the use of helicopters. The easiest to maintain would be the median version (good access to all sections of the pipeline), and the maritime version would be more difficult to maintain than the median one because of marshes.

Whatever the quality of the gas pipeline construction, however, the possibility of failure cannot be totally ruled out. Experience of pipeline maintenance under other terrain conditions suggests that the maritime version would be the least reliable, since a large section of the pipeline would run through an active corrosion medium (swamps). Here, as the pipeline ages, the probability of failure would increase. A similar assessment applied to the piedmont version, but for a different reason: experience of maintenance in mountain regions shows that there is a possibility of failure due to landslides and elimination of this is extremely difficult.

The most reliable is the median version where laying conditions are most favorable; it also has the best maintenance conditions which, in turn, increase its reliability.

5.4. Influence on the Environment

Based on this criterion, the maritime route is the most preferable passing through the marshy Kolkhida lowland area. The median route affects unique ancient forests and, to a greater degree than the two other versions, passes through agricultural land, citrus orchards, and tea plantations. Although the loss of this land would be temporary (for the period of construction), it would still be very undesirable.

The most undesirable route according to this criterion is the piedmont one. The cutting of terraces along mountain slopes would adversely affect the environment, and landslides could occur in consequence. Besides this, the construction of terraces would require greater amounts of land than are needed when laying pipelines on flat areas.

5.5. Connection with Regional Development Plan

The median and maritime routes would require much the same number of buildings to be demolished (69 and 61 respectively), but the piedmont version would be considerably worse (136). From the point of view of agricultural crop damage the piedmont version was again the worst (129 hectares) followed by the median version (102 hectares), and the maritime version (57 hectares). However, from the point of view of regional plans for the gas supply to potential consumers the median version was much better, so that this was the version favored by the local authorities.

5.6. Construction Conditions

According to this criterion, which is greatly dependent on the relief and other physical characteristics of the district, the median version has the best assessment. The maritime version was not so good and the piedmont version was much worse.

5.8. Population Safety

Existing standards for gas pipeline laying define necessary minimum distances from the pipeline to residential areas. Certainly, in the event of a pipeline failure, a gas leak resulting in fire risk can occur, but with the adopted working pressures and materials used in the pipeline itself, failure is very unlikely. Nevertheless, this possibility must still be considered, and here the maritime version is preferable, since it affects the smallest number of settlements, agricultural lands, and highways. The other two versions are approximately equivalent.

The analysis given above allows us to exclude the piedmont version from further consideration. The two other versions require additional analysis and comparison.

6. SELECTION PROCEDURES

Of the parties involved in the actual pipeline selection procedures, four major participants can be singled out. First, there is the customer organization, which determines the design task and performs pipeline maintenance; secondly, there is the organization that designs the pipeline; thirdly, any project has to be agreed upon with the regional authorities, which represent the interests of the local population; and, finally, the route selection is influenced by the contractor who will actually construct the pipeline.

When comparing the routes each participant in the selection process is primarily concerned with a definite subset of the given criteria. For example, the project organization draws attention to criteria C, C_1, C_2, IN, R, and S; regional authorities are concerned with criteria RP, IN, S, R, and C_2; and the customer is naturally interested in criteria C, M, R, and S. Finally, the contractor gives primary consideration to criteria T_{min} and S.

The selection procedures adopted are as follows. The project organization analyzes all possible pipeline routes. Using the initial basic outlines, the route direction in each version is then specified as that minimizing the presented costs. Then the project organization selects a version and transfers this proposal together with information about all the other versions to the customer and then to the regional authorities for approval. The contractor's representatives also take part in these discussions. In this example the project organization preferred the maritime version. When considering the various versions, the regional authorities pointed out the comparison between the far superior evaluations of the median version on criteria C_2, RP and R and the "best" evaluations of the maritime version on criteria IN and S. During the analysis the regional authorities asked the customer and the project organization to find new technical solutions to improve the evaluations of the median version on criteria S and IN in order to bring them nearer to the maritime version evaulation. As a result of investigations towards this end the project organization suggested the possibility of cutting down the guarding zone, combined with an increase in reliability effected by increasing the thickness of the pipe wall. It was found that with such an improvement the number of buildings requiring demolition would be considerably reduced and the presented costs of the median and maritime versions would become closer, despite the increase in the amount of metal required and in the cost of the pipeline. In Table 1 evaluations of the versions after incorporating this improvement are given.

With these improvements, all the participants in the selection process selected the medium version as the most acceptable, and so this version was chosen.

The example given above is typical in gas pipeline route selection. Each active participant in the procedure is at first guided by his own subset of criteria, working through from the more to the less important ones. This is characteristic of a satisfactory decision search according to Simon. We must point out that usually no single version is superior on all criteria; it is almost always necessary to look for a compromise. A typical feature of an actual comparison process is a series of attempts to revise some of the versions, in order to improve thier assessments on particular criteria.

Table 1.

Item	Criterion	Designation	Order of preference		
			Maritime	Median	Piedmont
(1)	Presented costs (million roubles)	C	8.9	9.5	10.8
(1A)	Cost of laying the main route (million roubles)	C_1	31	40	46
(1B)	Cost of laying prospective pipeline branches to consumers (million roubles, minimum)	C_2	9.5	5	5
(2)	Construction time	T_{min}	Second best	Best	Worst
(3)	Convenience of maintenance	M	Inferior	By far the best	Inferior
(4)	Reliability of maintenance	R	Best	Inferior	By far the worst
(5)	Influence on the environment	IN	Best	Inferior	By far the worst
(6)	Connection with regional development plans	RP	Second best	By far the best	Worst
(7)	Construction conditions	B	Second best	Best	By far the worst
(8)	Population safety	S	Best	Inferior	Inferior

7. GAS PIPELINE ROUTE SELECTION AND DECISION MAKING METHODS

From the point of view of decision theory the task considered in this paper constitutes a decision making problem with several parties (organizations) making decisions and evaluating various decision possibilities on a number of criteria (some criteria may be common to several versions). The number of versions considered is usually not large (2-5), but the criteria considered may be more numerous (6-12), and these are usually qualitative. It is important to realize that each gas pipeline is unique; therefore accurate quantitative data are not available, although experienced experts can give estimates for comparison purposes.

Gas pipeline route selection also represents an example of the problem of decision making involving a definite (although perhaps very insignificant) possibility of failure. The question arises as to whether it is possible to estimate the probability of failure (small or large), and the possible number of victims and amount of damage that would be caused by such a failure, etc.

Analysis of actual decision making procedures shows that such estimates are usually given in a wordy form. Naturally, these estimates are based on past experience, of breaches in normal pipeline operation as well as the conditions where such breaches occur. When selecting a route the designers and customers try to avoid such conditions, to take additional measures to increase reliability (e.g., by increasing the number of pipelines) and safety (perhaps changes in route direction) to an acceptable level by various amendments to the original plan. Usually these estimates are lengthy descriptions of various past incidents; certainly existing information will affect the estimates.

The unique character of each route and the lack of available statistics makes it impossible to obtain objective quantitative estimates. Subjective quantitative estimates given by experts are unreliable for the following reasons:

- (i) experts are not used to giving parameter estimates (except for cost) in quantitative form;
- (ii) it is difficult to separte the expert's past experience from his understanding of the peculiarities of the system he is investigating.

The great expenditure involved in pipeline construction makes the problem of perfection of the selection procedure very important. The question then arises of what can be achieved by utilizing the various methods of decision making to obtain a solution to the given problem and which methods are appropriate when the task peculiarities are taken into account.

Naturally, this question may be considered at two levels: that of the individual decision maker; and that of the decision making group. Because of the peculiarities of this decision making problem involving several decision makers, it is our opinion that the methods for determining the common utility of alternative decisions are often unsuitable, e.g., expected utility (Fishburn 1970), and multi-attribute utility theory (MacCrimmon and Sin 1974). There are two factors that hamper the use of these methods:

- (i) the difficulty in obtaining information in a quantitative form;
- (ii) the small number of decision alternatives; these make the procedure of comparison less labor-consuming than measuring the utility of each of them.

With a small number of alternatives, trade-off analyses are more appropriate (Keeney and Raiffa 1976); these enable qualitative estimates of alternatives to be used, especially in comparisons of their "character". Selection of the best version is performed through binary comparisons of the various versions, in which estimates for separate criteria are compared.

Studies of such procedures have shown the possibility of intransitiveness appearing (Tversky 1969). These studies have also shown that when using binary comparisons of versions involving estimates for numerous criteria, people tend to utilize simplified heuristics, of which the following should be mentioned: (a) consideration of criteria in turn; (b) disregarding of some criteria; (c) simple calculation of the number of criteria for which one version is found to be superior to another. Although such simplified heuristics are of great value, in some cases they can lead to intransitiveness. However, with the small number of versions considered, this possibility is not great, so that cases where nontransitiveness appear can be detected and eliminated fairly easily. Data from descriptive studies show which requirements have to be met by trade-off analyses in order to avoid the distortions induced by the limits of human cognitive ability in multi-dimensional information processing.

In order to avoid undesirable heuristics it is necessary for decision makers to consider information in sections, for instance, by comparing conflicting data on only two criteria at a time (Larichev and Kozhukharov 1979). Also, if the comparison system is biased then it is desirable for the decision makers to consider using a different one. It is also desirable to hasten the comparison process by agreeing quickly on the necessity of a compromise between competing aims. Comparison procedures should include methods of checking information even where there appears to be no discrepancy.

Possible methods of improvement of the procedures for perference correlation should be investigated. The primary efforts in the elaboration of route alternatives are made by the designers, who are also the first to carry out comparisons. From the point of view of the rationality of the whole process of decision making, it is desirable for the organization designer to take into consideration the whole set of estimation criteria for the various alternatives, together with any ideas put forward by other participants. In the final analysis, the decision maker (or designer) introduces his own preferences into the comparisons even when taking into account all the criteria. However, preliminary estimation of the viewpoints of the other decision makers will help the designer to control better the development of a proposed version. Anticipating objections, a decision maker can show in advance all the negative consequences of the selection of other versions, and this improves the selection process.

8. CONCLUSION

Many problems of decision making where risk and uncertainty are involved arise in the world around us, where a possibility of major failure exists; particularly problems of natural gas output, transport, liquefaction, and storage. Any possibility of real improvement in the processes of decision making where risk and uncertainty are involved should be used. In an attempt to find such a possibility, certain methods can be applied to elaborate the decision making methods. A rational basis for such methods is a compromise between descriptive and normative approaches. Knowledge of information available and human limitations should form the basis of normative decision making methods.

REFERENCES

Belousov, V.D., Bleikher, E.M., Nemudrov, E.M., Yufin, V.A., and Iakovlev, E.I. (1978) Gas and Oil Pipeline Transport. Moscow: Nedra.

Fishburn, P.C. (1979) Utility Theory for Decision Making. New York: Wiley.

Goncharov, V.I., and Oseredjko, Y.S. (1977) Questions of Systems Approach to Gas Pipeline Systems Designing with the Help of Computers. Kiev: Znanie Society of Ukrainian SSR.

Keeney, R., and Raiffa, H. (1976) Decisions with Multiple Objectives: Preference and Value Tradeoffs. New York: Wiley.

Larichev, O.I., and Kozhukharov, A.N. (1979) Multiple Criteria Assignment Problem: Combining the Collective Criterion with Indivudal Preferences. Mathématiques et Sciences Humaines 68:63-77.

MacCrimmon, K.R., and J.K. Sin (1974) Making trade-offs. Decision Sciences 5.

Tversky, A. (1969) Intransitivity of preferences. Psychological Review 76:31-48.

Worldwide Standards and Practices for Siting LNG and LPG Facilities: A Comparison

Elisabeth M. Drake and Marian H. Long

Risk Analysis Group, Arthur D. Little Inc., Cambridge, Massachusetts, USA

1. INTRODUCTION

Some perspective on the present codes and standards used for the siting, design, and operation of hazardous material facilities can be gained by looking at some of the factors influencing their evolution. Industry has a basic need for some standards that will, as a minimum,

(i) ensure that the facility will operate reliably, thus protecting the investment in the facility and the profits from its operation;
(ii) prevent less responsible members of industry from operating in a manner that could give the industry a bad name;
(iii) provide an environment in which employees can work without fear; and
(iv) maintain a good corporate image.

Since the trade-offs between safety and profitability are difficult to make and since, after a reasonable safety level is achieved, incremental safety becomes increasingly expensive, members of industry sectors tend to cooperate in establishing minimum standards. In the USA, groups such as the National Fire Protection Association (NFPA) and the American Petroleum Association (API) formed committees comprised of member representatives to establish minimum industry standards. These groups did a good job considering the difficulty of their task, and remained active to update standards as knowledge increased and new technology developed.

For example, the oldest and most widely employed liquefied petroleum gas (LPG)—or propane/butane mixtures—standards in the world are NFPA 58, *Storage and Handling of Liquefied Petroleum Gases*, and NFPA 59, *Storage and Handling of Liquefied Petroleum Gases at Utility Gas Plants*. Standard No. 58 for LPG was first adopted by the NFPA in 1932. Between 1932 and 1940, the rapidly growing use of LPG led to separate NFPA standards for different LPG applications. In 1940, each of these was combined into the single standard No. 58. The current version is the 1979 edition. Standard No. 59, which applies to LPG at utility gas plants, was originally developed in 1949 and is the product of a cooperative effort between the American Gas Association (AGA) and the NFPA.

However, people are imperfect and committees tend to operate in a somewhat random manner, as all of us who have served on committees know. Thus, unfortunately, some changes in codes, standards, and practices are due to the occurrence of unexpected accidents. Accidents also occur randomly, so it is not surprising that industry standards are somewhat uneven in the levels of safety provided. A further complication is the fact that larger companies tend to control the highest level of expertise and often have different attitudes from smaller companies who are less sophisticated and may bend industry standards either inadvertently or through a desire to stay in business. The possibility also exists that larger companies can hurt the competitive position of responsible smaller companies by enacting overly stringent industry standards.

---

*The views expressed in this paper are based on personal experiences and may not be accurate in some details, although the authors believe them to be generally valid.

If the industry picture is not complex enough, consider the situation when local, state, and federal governments get involved. Industry operates to protect its various self-interests, and governmental groups do likewise. The public, which has the least knowledge about the technology involved, is concerned about its safety and is distrustful of both industry and politicians. Finally, the press and TV dwell on sensationalism to promote their own interests. This *pot pourri* of interest groups and circumstances plays an important role in the evolution of the codes and standards for hazardous materials.

2. USE AND PROPERTIES OF LNG AND LPG

As natural gas became an increasingly desirable fuel in the USA, a pipeline system was developed between producing and consuming areas. However, in winter, demand was much greater than it was in summer, and an economic design of pipeline systems mitigated against design for peak capacity. Thus pipelines were designed for average winter use loads and other means of obtaining supplemental fuel for peak loads were found. These included storage of gas in large atmospheric pressure holders or use of propane-air mixtures to the extent that could be tolerated without affecting the burning characteristics of the mixture. Another option was to store liquefied natural gas (LNG) during the summer (although not as efficient as water, which is about 1000 times more dense in its liquid phase than vapor, LNG is about 600 times more dense than natural gas vapor) in well insulated storage tanks operating at slightly above atmospheric pressure and a temperature of -162°C.

At atmospheric pressure, propane liquefies around -40°C and can be held at ambient temperature as a liquid at pressures of around 10 bars (the pressure increases as ambient temperature increases). Table 1 shows typical properties of LNG, LPG, and similar fuels. Most of us are familiar with pressurized LPG which is widely used as a fuel by people in areas not served by natural gas distribution systems, and for cooking and heating in recreational vehicles. LPG can also be used as an alternative fuel for automobiles.

Table 1. Typical properties of some fuels.

	Methane	Ethylene	Ethane	Propane	n-butane	Typical light gasoline
Chemical formula	CH_4	C_2H_4	C_2H_6	C_3H_8	C_4H_{10}	C_5+
Molecular weight	16	28	30	44	58	~86
Atmospheric boiling temperature (°C)	-162	-104	-89	-42	0	47
Liquid density at atmospheric boiling point (kg/m^3)	427	567	544	583	602	656
Vapor density at atmospheric boiling point (kg/m^3)	1.75	2.09	2.06	2.43	2.71	3.0
Vapor pressure (atm) at 50°C	n.a.	n.a.	~50	18	5	~1
Lower and upper flammable limits in air (mol.%)	5-15	3.1-32	3.0-12.5	2.1-9.5	1.9-8.5	1-7

Typically, "LNG" refers to mixtures of the fuels in Table 1 where methane is the primary component; "LPG" generally refers to mixtures that are primarily propane and/or butane. Ethane and ethylene are often handled separately since ethylene is an important chemical feedstock.

These hydrocarbon gases are often found in association with oil. Oil production in more remote areas often involves the flaring of large quantities of associated gas. As the need for energy grew, incentives developed for utilizing this gas. Plans for liquefaction of gas for marine transport to consuming countries developed; large projects involving LNG shipments to Europe, Japan, and the US evolved; and shipment of refrigerated LPG became more common. While uninsulated, pressurized storage of LPG is feasible for relatively small quantities of fuel, quantities in excess of a few thousand cubic meters (liquid) are more economically transported as a low-temperature liquid at just above atmospheric pressure. Most large marine transportation projects involve refrigerated LPG or LNG.

3. DEVELOPMENT OF LNG REGULATIONS IN THE US

In 1941, the East Ohio Gas Company tried a pioneering experiment. They built a LNG peak-shaving facility in Cleveland, Ohio. The project was successful and additional tanks were built—in spite of wartime shortages of nickel, a key component in the steel alloy required for the low storage temperature. In 1944, however, one of the newest tanks failed due to brittle metallurgical fracture. The spilling cold liquid overflowed dikes designed for slow leaks, spread rapidly into the adjacent street about 20 m away, boiled rapidly, and vaporized. The vapors mixed with air and soon reached a source of ignition causing a fire in which 128 people died—most of them gas company workers in a building near the storage tank. This disaster aroused great concern, both from the public, government groups, and the gas industry eager to try the new technology. Nearly two decades passed before this form of natural gas system peak-shaving was tried again.

In the interim, the National Aeronautics and Space Administration (NASA) and the Department of Defense (DOD) became interested in cryogenic (low-temperature) rocket fuels and oxidizers. Much research was conducted into metals capable of containing liquid hydrogen (normal boiling point -253°C) and liquid oxygen (normal boilding point -184°C). This work provided information on improved cryogenic materials and procedures for safe handling of cryogens. While the storage of liquid oxygen required stainless steel (18 percent nickel), it was found that warmer LNG storage required 9 percent nickel steel—the steel at Cleveland had been about 3.5 percent nickel.

The other hazardous materials industries continued to operate along the guidelines of industry standards. With the increase in trucking and pipelining, as well as the deterioration of the rail system, the number of hazardous material transportation accidents increased. Concern tended to shift toward public sector regulatory control. The Natural Gas Pipeline Safety Act of 1968 gave the US Department of Transportation (DOT) responsibility for natural gas pipelining and facilities attached to the system. Thus, DOT has control over LNG facilities connected to the natural gas pipeline networks. The US Coast Guard (USCG) was given jurisdiction over all marine shipments including those involving a wide variety of toxic, flammable, and explosive materials.

In the 1960s, the need for peak-shaving gas grew to the point where gas companies again started constructing LNG facilities. For many years, LNG facilities were designed to meet the requirements of NFPA 59A. This was periodically revised and updated by its advisory committee, which included representatives of the gas industry, insurance companies, federal agencies, manufacturers and consulting firms. NFPA 59A was intended to be a minimum standard that would provide a basis for safe design of LNG peak-shaving plants, and recently it has been amplified to include some additional requirements for import terminals. In a number of respects, the NFPA 59A code is more stringent than codes covering the construction of facilities for a wide range of flammable and/or toxic hazardous materials.

A few state and local authorities with jurisdiction over LNG facilities adopted regulations which duplicated or amplified the requirements of NFPA 59A. Even before the adoption of the first version of NFPA 59A (1967 edition), the Massachusetts Department of Public Utilities issued regulations closely following the NFPA 59A draft code. Also, the New York City Fire Department (NYCFD) set requirements for LNG facilities built in New York City which significantly exceeded the NFPA 59A standards with regard to diking and buffer zone requirements for LNG storage systems. The NYCFD indicated that they felt that more stringent requirements were appropriate for LNG facilities sited in heavily congested areas.

In the early 1970s public and regulatory concern began to be voiced about LNG safety, but it is somewhat ironic that this concern arose in the aftermath of a tragic accident in Staten Island, New York, in 1972. The accident involved 40 construction workers who were making repairs on the inside of an empty LNG tank when ignition of flammable insulation in the tank occurred. While the accident did not compromise public safety, it nevertheless greatly amplified public and political concerns about LNG.

At the time of the accident, almost all the LNG facilities in the US were designed for peak-shaving purposes. The two exceptions were an export terminal in Alaska and a relatively small-scale import terminal in Massachusetts. However, several major import projects were in various stages of development and public anxiety turned to large LNG ships visiting major US ports to supply baseload terminals.

In response to public concern, regulatory activity began to increase. The Office of Pipeline Safety (now part of the Materials Transportation Bureau—MTB) in the US Department of Transportation (DOT) had responsibility for LNG facilities under the Natural Gas Pipeline Safety Act of 1968. In October 1972, DOT adopted NFPA 59A (1971 edition) as an interim federal standard and started to develop their own regulations. The 1972 edition was subsequently adopted. A report summarizing existing LNG regulations, the status of LNG safety research, issues of concern about LNG safety, and the applicability of risk management methods to LNG facilities was prepared by Arthur D. Little, Inc. for DOT in 1974. This report was intended to provide DOT with a database from which they could proceed in developing federal LNG safety standards. Probably because of the complexity and controversial nature of the issues as well as limited internal resources, the DOT Office of Pipeline Safety Operations (OPSO) did not publish the *Advance Notice of Proposed Rulemaking* (ANPRM) until April 1977. In the introductory material to the OPSO ANPRM, DOT stated that the notice was a step in the developmental process to produce permanent federal regulations and that:

> "Although this notice is not a proposal to amend the present standards, it contains a comprehensive set of draft regulations which are intended to serve as a basis for public comment and participation in identification of LNG safety problems and the development of appropriate regulatory solutions to those problems, considering all reasonable alternatives. Comments to this notice should assure that if a new Part 193 is adopted, it is founded on a broad source of information."

The OPSO ANPRM served its purpose in arousing comment and many groups and individuals submitted comments which in a number of cases identified technical errors or inconsistencies. Unfortunately, many state, local, and even international regulatory groups without much depth of technical expertise in LNG viewed this document as a new US federal regulation. They were further confused, on the one hand, by some industry groups who took a strong position against the OPSO ANPRM and argued that the existing NFPA 59A standards were completely satifactory, and on the other hand, by groups who claimed that the NFPA 59A code was an industry code and that did not adequately represent public concern for their own safety. This latter view was reinforced by varying degrees in two reports, one prepared by the General Accounting Office (GAO) and another by the Office of Technology Assessment (OTA). It was the GAO report which first publicly grouped LPG with LNG and gave these materials the general name of liquefied energy gases (LEG).

Regulatory groups at the state level started to respond to the circumstances. California, Massachusetts, and New York are of particular interest in providing an overview. In each of these states, a legislative action stimulated development of further regulations governing siting criteria and safety.

New York. The New York Department of Environmental Quality (NYDEQ) was given a lead role in establishing siting criteria, and published a draft regulation which essentially combined the NYCFD requirements and those of the OPSO ANPRM. The result was a voluminous and confusing document. Further action is still pending.

Massachusetts. An Energy Facilities Siting Council (EFSC) was established and was given prime responsibility for the siting of LNG facilities in the Commonwealth. The EFSC subsequently promulgated LNG facility siting guidelines, the requirements of which are such that it is unlikely that a new LNG facility will be built in Massachusetts. The EFSC has stated that their guidelines are an interim measure and that they planned to accept new federal regulations if and when such were promulgated. They have not yet done so.

California. The California Public Utilities Commission (CPUC) was directed to prepare LNG safety regulations as part of the LNG Terminal Act of 1977. A comprehensive set of regulations has been prepared, following the ANPRM format but differing substantially from the ANPRM in a number of aspects. These rules contain a particularly comprehensive set of seismic design requirements.

Meanwhile, back in Washington, DC, the US Coast Guard (USCG) was working on safety standards for waterfront LNG facilities under the authority of the Ports and Waterways Safety Act of 1972. In February 1978, the USCG and MTB (both part of the DOT) signed a memorandum of understanding clarifying the split in jurisdiction between the two agencies. In this agreement, MTB would have jurisdiction over the shore-based portion of waterfront facilities with the exception of fire protection and security systems which were to be the responsibility of the USCG.

In April 1979, MTB published a notice of proposed rulemaking and later published final rules. The rules relating to siting, design, and construction were promulgated on 30 January 1980 with existing facilities "grandfathered" unless they were significantly altered by increasing the storage capacity or by relocation. On 28 August 1980, additional changes in the siting, design, and construction standards were issued by MTB, improving and clarifying certain requirements, especially in the area of seismic design. New rules covering operations, maintenance, and security of LNG facilities were enacted on 23 October 1980, applicable to both new and existing facilities. An implementation period was allowed for existing facilities.

With the new US federal LNG regulations in force, some of the states and localities which developed their own rules are now in the process of reviewing them and deciding whether or not to rely on the federal standard.

4. LPG STANDARDS IN THE US

In the US, LPG ranks fourth in supplying the nation's energy needs with an annual consumption of about 70 million m^3. The consumers, who include homes, farms, individuals, businesses, and government groups, total approximately 60 million (1977).

LPG is easily stored, transported, and regulated, and is employed for a variety of tasks. The homeowner uses it to warm his home, heat water, and cook food. It is particularly practical in rural areas: on farms, it is used in crop drying, defoliation, pig breeding, and frost protection. In industry, it is often used for soldering, cutting, heat treating, and vulcanizing. It is valuable as an engine fuel as it has minimal emissions.

The distribution of LPG throughout the US involves more than 86 000 people, using 225 000 miles of cross-country pipelines, 25 000 transport and delivery trucks, 22 000 railroad tank cars, 250 primary storage facilities with a capacity of 7 billion gallons, a fleet of 370 barges and tankers, 8000 bulk storage and distribution points and 25 000 retail outlets. The LPG distributed is produced by more than 200 oil and gas companies (1977)*.

*Source: Energy Information Administration.

As indicated in the Introduction, LPG facilities in the US are currently constructed to meet NFPA standards which in turn reference design standards of the API and the American Society of Mechanical Engineers (ASME). In addition, requirements of local building codes must be met.

Traditionally, design against natural disasters at facilities for handling most flammable and toxic materials has been based on building code criteria. The building code relies on past observations of natural phenomena in particular localities. Often the observation period is in the range of 20-50 years with less precise information available for up to 100 years or so. The building codes typically include design conditions for wind, seismic, and snow loads. Because the design conditions based on limited historic data do occasionally occur, safety factors are incorporated in the design codes to provide additional margins of safety for these and other reasons. Considering the large number of hazardous material storage tanks in the US and the rarity with which failures of these tanks occur due to natural disasters, the building code criteria seem to be generally acceptable at present. However, the DOT has indicated future plans for developing federal safety regulations pertaining to LPG facilities.

5. COMPARISON OF LNG AND LPG STANDARDS IN THE US

At the present time, US federal LNG standards are considerably more stringent than standards for LPG (or a wide variety of other hazardous materials). For example, in the past decade, there has been a growing trend to characterize natural phenomena in probabilistic terms related to an expected frequency of occurrence of events of various levels of severity. The US Geological Survey (USGS) has issued a draft for a new seismic design criteria map for the US which may replace current building code seismic criteria. The USGS map presents seismic acceleration design levels in terms of percentages of gravitational acceleration. The levels are chosen so that at any location there is a 90 percent probability that the design level will not be exceeded in 50 years. This is equivalent to an event with an annual probability of about 2.1×10^{-3} per year, or an expected return interval of 475 years. In the accompanying text, the USGS comments that a more stringent criterion should be applied to the seismic design of critical structures like major dams or nuclear containment vessels. USGS mentions that a criterion of a 99.5 percent chance of not being exceeded in 50 years might be appropriate. This event has a 10^{-4} per year probability and an expected return interval of 9975 years.

While most petrochemical facilities, including LPG, are designed to existing building codes, storage tanks at LNG facilities must be designed for a 10 000 year return interval for winds or seismic events. A minimum design wind of 200 mph is to be used if data are not available to determine the 10 000 year wind speed. Thus, an LNG storage tank is required to withstand at least 200 mph winds, while LPG and other typical fuel storage tanks may be designed in accordance with a building code for 100 mph winds. In comparison, a nuclear containment vessel must be designed for 300 mph winds (or 360 mph in highly seismic zones) associated with an unusually severe tornado. The DOT LNG regulations are somewhat inconsistent in that design for flooding is only based on the worst occurrence in a 100-year period.

Another area of comparison relates to buffer zone requirements. The DOT regulations require diking of all LNG storage tanks with a separation distance between the dike edge and the property line equal to $f\sqrt{A}$, where A is the dike area and f ranges from 0.8 (minimum) to 3 if the adjacent area is a place of outdoor public assembly (beaches, playgrounds, theaters, etc.). The NFPA code for LNG uses a single factor of 0.8 for the buffer zone. For LPG, several standards apply to buffer zone requirements:

(i) NFPA 58 states the minimum distance that an above ground or an underground container may be installed from "important buildings" or an adjoining property line, but does not consider containers larger than $455 m^3$ "Important buildings" are not defined.

(ii) NFPA 59 details such distance information for nonrefrigerated and refrigerated above ground containers, respectively. For containers larger than 378m^3 capacity, the distance from the container to the nearest important building not associated with the LPG plant or to an active property line is 121m. Lesser distances are specified for smaller quantities.

(iii) API 2510, *Design and Construction of LP-Gas Installations at Marine and Pipeline Terminals, Natural Gas Processing Plants, Refineries, Petrochemical Plants, and Tank Farms*, states the minimum distance from an adjoining property to an LPG container for both refrigerated and unrefrigerated tanks. For the latter with a capacity above 2000m^3, the distance is 60m. For refrigerated tanks, the minimum distance from adjoining property and the tanks is a distance of 1½ times the tank diameter or 60m, whichever is less.

The standards for LPG do not address the question of potentially flammable vapor clouds crossing a property line, while the DOT LNG standards do. Design accidents are specified for which buffer zones must be provided to prevent flammable vapor (defined as 2.5 percent methane or half the lower flammable limit) from crossing land not under the control of the facility owner. The design accident involves the drainage of a full storage tank contents through the largest bottom piping penetration on the tank or, for tanks with all piping through the roof, a ten-minute spill at the maximum flow rate. A similar, but less severe, requirement is in the NFPA 59A, 1979 edition.

In the area of marine transport of LNG and refrigerated LPG, the USCG rules treat the two materials equivalently (except for cargo design temperatures), requiring double hulls and bottoms as well as special USCG controls for ships entering US ports.

6. LNG AND LPG STANDARDS OUTSIDE THE US

Most European countries and Japan have a procedure where the owner of a proposed hazardous facility (LNG, LPG, or various other flammable or toxic materials) must apply to a local or regional group for permission to proceed with construction. In the UK, an applicant provides information on a proposed project design and potential impacts—environmental, safety, and societal. Initially, the regional group will give outline planning permission after public hearings if the project seems desirable. Conditions may be attached to this planning permission which may impose further requirements such as an independent safety audit conducted to the satisfaction of the Health and Safety Executive. These conditions may also add some requirements which exceed normal industry practices and codes.

Countries like New Zealand and Australia tend to follow the UK practice of application to local authorities, public hearings, and planning permission. In Japan, a similar system also seems to be followed, with the local authority often assembling an advisory panel of experts to assist them in more complicated technical issues.

In September 1980, West Germany passed a chemicals act with the purpose of protecting man and the environment from the harmful effects of dangerous substances (defined as toxic, flammable, corrosive, explosive, carcinogenic, etc). A manufacturer of a new dangerous substance is required to notify the government at least 45 days before initial circulation of the product. (Parallel requirements are being developed among other member countries which are of the European Communities.) The government has the authority to prohibit or limit the use of substances creating undue risk to human life, health, or the environment. The Bundesrat has also authorized the development of regulations to cover occupational and public safety with respect to dangerous substances. Regulations covering marine transport do not vary very much worldwide, since the International Maritime Consultative Organization (IMCO) has adopted standards which are essentially the same as those of the USCG.

7. DETAILED REGULATIONS FOR LNG AND LPG

For LNG, most countries have followed US practice, generally adopting NFPA 59A as the basic standard. In some countries, such as the UK, NFPA was adapted to reference British standards in lieu of API and ASME standards.

Many countries are now pondering the US DOT LNG regulations; portions of these new regulations—notably the flammable vapor exclusion zone and seismic and wind design—exceed past practice and present industry and regulatory practice for a variety of other hazardous materials. For example, in Japan, LNG facilities are now designed for flammable vapor dispersion buffer zones, but a three minute design spill is used (as compared with 10 minutes in DOT) and vapors at the property must be diluted to a 5 percent concentration (2.5 percent in DOT). To the authors' knowledge, no major facility has yet been built or planned in the US which fully meets the new DOT LNG regulations.

There are several British codes and standards which govern LPG operations. In 1973, the Department of Employment published a Health and Safety at Work (HSW) code entitled *The Storage of Liquefied Petroleum Gas at Factories* which specifically addresses precautions for storage and handling, ignition source control and fire protection. In 1970, *Liquefied Flammable Gases Storage and Handling—Engineering Codes and Regulations* was published by Imperial Chemical Industries Ltd., and the Royal Society for the Prevention of Accidents (ROSPA). Addressed in these regulations are safety distances for location and spacing, ignition source control, pressure relief design, fire protection, and road, rail, and ship tanker loading and unloading. The Liquefied Petroleum Gas Safety Code, which is part 9 of the Institute of Petroleum Model Code of Safe Practice in the Petroleum Industry, was jointly prepared by the Institute of Petroleum, the Institution of Gas Engineers, and the Liquefied Petroleum Gas Industry Technical Committee. First published in 1967, it was last revised in 1975, and specifies storage tank design and location, fire protection, and rail tank car design. Other UK codes and standards pertaining to LPG include *Safety Recommendation IGE/SR/ 6—Liquefied Petroleum Gases* published by the Institute of Gas Engineers, and the *Code and Practice for the Storage of Liquefied Petroleum Gas at Fixed Installations* published by the Home Office.

The Standards Association of Australia has published Standard No. 1596—*Rules for the Storage and Handling of Liquefied Petroleum Gases*. First published in 1973, it was last reprinted in 1978, and specifies detailed regulations and requirements for both above ground and underground storage tanks, rail tank cars, and liquid transfer.

The New Zealand Fire Service has published an information guide, *The Safe Handling of Liquefied Petroleum Gas and Engineering Procedures*, which includes information on purging and ventilation, multi-mode transportation, cooling water application, container failures, the boiling liquid, expanding vapor explosion (BLEVE) operational procedures, fire prevention, and fire protection.

8. FURTHER EVOLUTION OF LNG AND LPG STANDARDS

Standards will continue to evolve as analysis or occurrence of accidents indicate weaknesses in the present requirements. An accident at the Cove Point, Maryland, LNG import terminal in October 1979 uncovered a deficiency in the design of certain seals in electrical systems. Changes are now being made in the NFPA codes and the National Electric Code to correct the deficiency.

Only a few major accidents have related to the extensive worldwide operations involving LPG energy storage and transportation—the most serious involved rail or road transport accidents. The US Department of Transportation is now requiring the use of head shields on LPG rail tank cars to prevent coupler punctures during derailments. In addition, LPG rail cars are now being coated with a fire-protective material to reduce the likelihood of "BLEVE-type" accidents, which usually occur after an LPG tank has been immersed in a fire for several minutes or more.

Following a refrigerated LPG tank failure in Qatar, a careful investigation of the accident was conducted and several changes were made in the draft British

Institute of Petroleum code specifications to which the tank was built. One of the changes brought metallurgical testing requirements into conformance to the somewhat more stringent US API and ASTM codes. In addition, we understand the British code now requires 100 percent hydrostatic testing of a refrigerated LPG storage tank before it is commissioned. This latter requirement is more stringent than that of the API code, which allows a partial hydrostatic test if the tank foundation (as designed for normal operating stress loadings) cannot take the loadings for a 100 percent hydrostatic test. The API code is currently being revised for LPG.

In the recent past, Shell International has proposed a double-integrity containment system for refrigerated LPG and LNG storage tanks in areas where a serious tank problem has potential to threaten public safety. The secondary containment system would be designed to withstand catastrophic failure of the primary container. There are, however, differences of opinion within the industry as to the need for designing the storage system to withstand a catastrophic failure of the primary containment vessel. This controversy is centered on the ability to be assured that the material of the primary containment vessel has crack arrest properties, which would preclude the possibility of the vessel unzippering. Work on the metallurgical properties of low-temperature materials is currently being conducted in order to resolve these questions.

While it is rare that regulations are ever made less stringent, it is clear that many of the present US regulations covering LNG are inconsistent—in terms of actual hazards—with rules for LPG and other hazardous materials. After considering the problems and deficiencies in previous attempts to assess economic impacts and benefits of new or proposed regulations, the question of how this may be done better arises. This is a difficult problem and one that is properly in the province of the regulatory agencies. It seems reasonable that new regulations should be developed with due regard to the fact that consumers are paying the costs of making risks acceptably low to abuttors of hazardous material facilities. We cannot afford zero risk and, in fact, do not require zero risk for a wide range of activities and exposures which we accept routinely. The public is now becoming increasingly aware of a whole spectrum of risks produced by our industrial sector, including problems with pollutants, carcinogens, hazardous waste disposal, toxic and flammable materials handling, etc. It will be costly to upgrade public safety, and we should spend our money wisely. I suggest that a decision maker might attempt to assess present levels (including ranges) of risks to which the public is exposed. Presumably the public is concerned that these levels are not adequate in a number of areas. Thus, as a society, we should attempt to identify risks that are currently significantly higher than the accepted range and attempt to reduce such risks through corrective measures which might include regulatory or industry actions, provision of incentives or penalties, etc. In addition, our standard of living and economic status is high enough that we can probably afford to push general levels of safety higher than they are at present.

In our opinion, it is a mistake to attempt to develop regulations on the premise that there is some remote potential for catastrophe, if the probability of the remote event is not also considered. Also, in focusing on the extremely rare event the control of some lesser but more likely risks may be overlooked. We are very aware of the large uncertainties in setting probability levels for very rate events. Nevertheless, such estimates—along with estimates of uncertainties—are useful tools in the decision process.

Whether they formally quantify probabilities or not, the regulatory decision makers are already estimating risks intuitively when maximum credible or design accidents are established. Only by comparing risks can decision makers determine whether a need for more stringent measures exists. If the risk levels seem to be too high, risk reduction measures and costs should be evaluated. Conversely, it is hoped that regulatory decision makers will also have the courage to amend, or adopt in modified form, regulations which are excessively stringent and increase consumer costs unnecessarily.

Since the existing discrepancies in regulations are probably due mostly to perceptions of risk rather than to true relative hazards, both the public and its representatives will have to be made more knowledgeable about hazardous material risks.

This is difficult to do without alarming people about hazards of which they might not be aware, but probably will be rquired if society is to find a reasonable balance between safety and costs to consumers.

Further, it seems that different localities or countries may have different existing background risks and different attitudes toward acceptable safety. At the most local level, there is the danger that each town may ban all hazardous facilities but hope that they will be located nearby so their economic benefits may be enjoyed. Thus, the approach of decision making in a national or regional forum appears to be the best solution.

Chapter 3

INSTITUTIONAL ASPECTS OF RISK

Risks and the Waning of Compromise in Politics

Volker Ronge

Infratest Sozialforschung, Schluderstrasse 4, D-8000 München 19, FRG

1. INTRODUCTION

My so far modest share in the field of risk research (which is by no means something clearly defined and homogeneous) has been to introduce some basic and, to my mind, essential information about political, economic, and social aspects underlying risk generation and risk taking, decisions, and reflections on risk (Ronge 1980). I held it necessary to remind risk researchers that these "fundamentals" of their analyses were still mainly technical in character at that time. This introduction of sociological knowldge was intended to promote risk reflections and decisions, i.e., to make them more realistic and effective. My present intention is different, and at first it will appear to be more external and less helpful. I am going to bring into risk debates—which, I suspect, have integrated some sociological thinking meanwhile—some concrete information about society which, to my mind, must affect risk decision making in the first instance, and risk research in the second.

In other words, my first contact with risk research was determined by the supposition that, with respect to single-risk studies, it was lacking sociological knowledge and reflection in general. My present intervention is based on the belief that drastic changes are occurring within society (i.e., Western, industrialized, capitalist countries with a special focus on the FRG) which affect, or should affect, risk decision making, and risk analysis as such, i.e., as a scientific undertaking that aims to rationalize social processes.

Although this argument must appear to be negative at first sight, in fact it is not. I admit that my arguments will not lead to immediate improvements in risk analyses and decisions, yet they may be useful in explaining a little of what I see currently happening to risk research, namely the experience and frustrating feeling of being unable to promote effectively risk decision making. I shall therefore address self-reflections and self-consciousness in risk research which in the long run will also affect its output and impact.

To put my message more concisely, the same social developments that once led to the need for, and the "invention" of, risk research may—it is still taking place—function as barriers to the immediate "execution" of scientific risk knowledge, i.e., its transfer into political decisions with the result of improving society's risk taking. Risk research (necessarily) aims at being, or at least supporting, the master of social processes, but it is in fact their victim. Only an external perspective can reveal this dilemma.

2. FROM TECHNICAL TO ECONOMIC RISK ANALYSIS

Let me start with some sketchy remarks on the recent development of risk research. That questions of risk (or safety) are to be treated in terms of decision making has recently become widely accepted in the field of risk research. Cost-benefit analysis is presented as an adequate tool to reduce risk of all kinds (cf. Conrad and Kresbach-Gnath 1980, p55). The fact that there are limited resources that can be spent for alternative purposes thus leads us to the point where we have to consider and compare the marginal costs. In business administration, for example, there are intensive discussions about how to integrate the costs of risk analysis into the overall cost-benefit analysis of investment (Schindel 1978). This can be expected to become a necessary consequence for every kind of risk analysis.

This development from a purely natural science perspective—with its interest in comparative probabilities of accidents*—towards an economic view of risks I regard as an essential step forward. It is, however, only one step in the right direction. What is missing still is the systematic consideration of these peculiarities that work in social or political decision making.** Economic rationality is far from being a social one; on the contrary, it is adequate only for one of society's subsystems (the economy)—and even this is in dispute. From a societal point of view, economic rationality is an inadequate abstraction, since it

(i) takes certain ends (such as long-term optimization of profits) for granted that are by no means common to everyone in society, or to society as a whole;
(ii) presupposes one particular medium of "social exchange" (money) that is by no means the only existing within society;
(iii) connects ends and means in a very straightforward way, thus ignoring a lot of (empirical) social obstacles which consequently, but unjustly, appear to be "irrational" from an economic point of view;
(iv) presupposes a coherent (normatively integrated) unit or sphere of action (the firm) for which decisions have to be made;
(v) tacitly functions under favorable external conditions (social costs externalized to society).

3. FROM ECONOMIC TO SOCIOLOGICAL RISK ANALYSIS

Let me demonstrate the transition from economic to sociological analysis by quoting a statement from a recent Norwegian publication on risk research (Hovden 1980, p9):

> "We can <u>reduce the risk</u> in any given sector provided we are prepared to <u>pay the cost</u>."

I have underlined those parts of the sentence which lead considerations to cost-benefit analysis and thus to economics. Just a small shift of emphasis is sufficient to signify the next step necessary:

> "<u>We</u> can reduce the risk in any given sector provided <u>we</u> are prepared to pay the cost."

"We" is nothing but a metaphor for "society", yet to use the word "we" is at the same time misleading insofar as it covers complex social processes that normally do not result in unanimous opinions, interests, and decisions, but in controversies and—possibly—compromises, with (groups of) winners and losers. Society is far from being a homogeneous, coherent "actor"—as would implicitly be taken for granted

---

*"From the technical perspective, technological risk is some probability distribution over sets of negative effects" (Lathrop 1981, p4). "However, there exists a concept of risk that considers both probability of failures and consequences of failures, and thus seems to be superior to the other concepts used" (Mandl and Lathrop 1981, p2). See also Conrad and Krebsbach-Gnath 1980, p50.

**I do not go into that sidestream of risk research that focuses on psychological processes (Conrad and Krebsbach-Gnath 1980, p57). It is obvious that these processes are severe restrictions to the model of *homo economicus*. Yet this is a general problem of every kind of decision, not specific to risk decisions. Moreoever, I restrict myself to the macro-level which has its own specific "laws"—however micro-processes function.

under the economic paradigm. It would therefore be misleading to look for generally "acceptable" levels of safety (or risk) i.e., levels for all members/groups of society, in all situations, or for all issues, equally.

> "The problem is that risk levels cannot be universally determined, but have to be specified for each hazardous activity and must be broken down to apply to the various categories of people involved. Standards of safety vary" (Hovden 1980, p26).

Put a little more rigidly: there is no *acceptability* of risks which could be found out by risk research, but there are *accepted* risks. This is the field which sociological analyses of risk have to deal with. It is the sociological approach that prevents us from regarding social processes as irrational.

4. TWO CONCEPTS OF SOCIOLOGY IN FRONT OF RISK

To simplify, there are two diverging points of reference for approaching the risk issue in sociological terms. The focus of analysis can be

(i) perception and activities of individuals summed up to some kind of mass aggregate called society*, or

(ii) the complex political process of forming society's "joint" will and action out of the individual's (or group's) perceptions, interests, and activities—mediated by structures—which is a decision making process implying controversies, bargaining, and (possibly) compromises.

I shall refrain from a detailed description and evaluaticn of these two approaches which, after all, provoke different methods of analysis**. Nevertheless, I prefer the latter. Sociology, to my mind, must be concerned with how perceptions and actions of individuals and groups are transformed into those of society as a whole. This means it must focus on the procedures which form social decisions.

5. FROM RISK ACCEPTABILITY/ACCEPTANCE TO LABILE CONSENSUS

There is some indication of insufficiency lying in the mere—however progressive—amplification of risk research to the issue of social acceptability or even acceptance. The FRG Minister of the Interior, Herr G. Baum, recently stated:

> "The question of acceptance has turned into the question of the scope of consensus about values and priorities" (Baum 1981, p31, my translation)***.

I shall follow up this statement, for I believe that there is some reason and truth in it. Before going into that in more detail, however, let me sum up the development of the dominant interests and paradigms of risk research according to the brief overview given up to now.

---

*This is the field of survey research. A critical review of surveys on nuclear power and corresponding risks is given by Heller (1980).

**I approve, however, of Hovden's argument (1980, p21) that "the risk perceptions of individuals and of society as a whole must be treated separately". They *are* something different, I would say.

***Roughly the same is said by Conrad and Krebsbach-Gnath (1980, p63) in their review of risk research literature: "Risk research turns into research on socio-political mechanisms" (my translation).

Abstract, possibly comparative calculation of risks; scenarios of failure consequences	→	Investigation into the acceptability/acceptance of risks by individuals, groups (of individuals), or *Homo economicus*	→	Research on consensus prerequisites for the politics of risk taking and distribution

The switch to the question of acceptability/acceptance was made some time ago, but at present, risk researchers working on this question feel increasingly frustrated by experiencing a growing and not easily understandable resistance of people to any kind of risk*. This leads me to go one step further into the abstract, into the problem of social consensus. Consensus is more abstract than acceptance in that it is directed towards (and withdrawn from) political activity in general, not towards (and from) concrete actions, e.g., some bearing some risk. Now, what is wrong with consensus today?

I shall start with the norms and functional prerequisites of the political system, and then turn to factual developments. The presupposition of Western democratic political procedures within society implies

— fundamental consensus on the political system, i.e., institutions and procedures,
— renunciation of illegal power leaving the "monopoly of physical power" to state institutions,
— acceptance of democratic decisions, i.e., the output of the political system, without claiming another authority (e.g., God, or truth) to be a serious competitor.

The crucial test for the acceptance of the political system and democratic procedures, or for the existence of consensus, is whether those who are the losers in certain political decision making processes will nevertheless accept its output. Frustrations are then turned into "official" channels, i.e., they become new inputs in turn.

Under these circumstances there is no need at all to talk separately of risks, as we can assume that nearly all activities within society—and therefore political decisions as well—entail certain tangible risks (Hovden 1980, p36). Risk is, so to speak, something "normal"**, and its social distribution is enacted in the ordinary course of political decision making—which may be assisted by scientific advice, one separate branch of which may be risk analysis. The social acceptance of certain risks is thus guaranteed by ordinary political decision making itself.

This argument from the opposite point of view leads me to suspect*** that the separate discussion and treatment of risks which has come about during the last

*One indicator of this frustration is the increasing interest of risk researchers in the processes and "laws" of political praxis, decision making in particular, which (unjustly) appears to be the main barrier to the enforcement of their risk knowledge.

**This is addressed in the report of a FRG commission on future nuclear power policy (Enquete-Kommission *Zukünftige Kernenergie-Politik*, Bundestags-Drucksache 8/4341, 27 June 1980): "Risks to man and the environment are concomitant to every energy system" (p139; my translation). The same idea is expressed by an official information booklet of the government of Baden-Württemberg on the issue of defense against catastrophes (Versorgung-Sicherheit-Katastrophenschutz, in: *Sicherheit in Chemie und Umwelt*, No. 1/1981, p19): "We are all permanently living under natural as well as technical risks. Thus all kinds of energy supply are bearing risks of different types" (my translation).

***Another line of argument is possible, namely that the separate discussion of risks indicates a new dimension or size of risks. The atomic energy issue could be seen in this respect—yet only this issue, I suspect. Even this is disputed, however, by the hypothesis that the nuclear power issue is only a "symbol" of quite different problems. This idea is particularly promoted by K.-M. Meyer-Abich (in:

delegation of power and authority to established political institutions (parliaments, parties, governments) is being questioned. The unquestioning acceptance of political decisions is decreasing. The function of, and competence for, decision making slips over from legislature and administration to the judiciary.

Reacting to these developments, political scientists have already begun to change their traditional view of policy making; new paradigms of politics are being discussed (Raschke 1980).

(1) Contrary to assumptions widely held during the 1950s and 1960s that we have approached an era without ideologies, there is a growing awareness of (i) fundamentally and rapidly changing values towards more ethical and moral ones (Inglehart 1977)*; (ii) a diminishing reluctance of individuals to use physical power (illegal power, which, however, is held to be legitimate) in order to support their interests.

(2) The relatively sharp separation of society and the political (decision making) system which has dominated political science thinking for a long time is being questioned. The fundamental consensus in society which meant agreement on certain political rules, institutions, and processes, and which led to an identificiation of legality and legitimacy, must perhaps be replaced by a concept that is supposed to be more adequate for the present status of society; a situation in which there is only a limited exchange between political authorities and citizens. In the new concept apathy is no longer regarded as a functional virtue, as it has been traditionally, and citizens are being looked at as potential political actors apart from official political procedures. The population (or certain parts of it, at least) accept the rules of policy making—or hold consensus—only insofar and as long as they expect them to be not unfavorable to their interests. This does not imply that unpleasant political outcomes are generally not accepted (even though this may occur), but a structural deprivation by political decisions is rejected. Therefore the political system has to face the increasing possibility of people reacting in an "unconventional" way (Kaase 1976).

8. WANING OF COMPROMISE—SOME EMPIRICAL REMARKS

The empirical proof of the existence as well as the character of the new social movements is still not available. My interpretation, however, cannot be very far from the truth when even a sensible politician like FRG Minister Baum speaks of "alarming symptoms in the present kinds of handling and settling social conflicts—brutalities in the course of demonstrations, transgressions, riots..." (Baum 1981, p13; my translation). The same description of the facts is implicit in the statement of FRG Chancellor Helmut Schmidt:

> "I want to call upon the young people in our country ... to learn, and be willing to accept social consensus, agreements and compromise. Ability of consensus and compromise is an underrated necessity in Germany today." (Schmidt, Lernziel: Bereitschaft zum gesellschaftlichen Konsens, in: *Bulletin des Bundespresseamts*, No. 62/1981, p528; my translation.)

---

*According to Schissler (1979, p356) a growing "individualization" and selfishness of individuals can be observed and, furthermore, a diminishing orientation towards supra-individual values and social responsibility. Following Inglehart (1979, pp 283-284) a clash is likely between an increasing number of people holding post-materialist values and socio-political "environment" which is still materialist more or less. It seems probable that the post-materialists will attempt to introduce radical and far-reaching changes within society and, in order to reach this aim, will be prepared to use means that are destructive and provocative to the elite too.

As this paper is not intended to be an empirical one, I am restricting myself to some sketchy remarks; detailed analysis, specific to single countries, would be necessary to give more validity to my assertions*.

(i) In West Germany (and elsewhere) we are experiencing a social movement—of young people in particular—in which all the (analytical) aspects mentioned above are intermingled. Not by accident is the movement called "alternative culture"—not "politics"—thus indicating the general and total difference of these new standards and activities from traditional behavior, including political activity, in Western societies. This (factual) movement can be nicely characterized by one of its own slogans: "No authority to nobody". As a recent paper by the German Social Democratic Party stated, this movement is not apolitical, but has a different understanding of politics:

"...from the delegation of social and also individual problem-solving competence to large organizations the structures of which are too fossilized, and prevent people from participation—towards self-responsibility and personal initiative" (Thesenpapier "SPD und Jugendprotest" der Arbeitsgruppe Jugend beim SPD-Parteivorstand, in: *Vorwärts* Nr. 27/1981, p19; my translation**).

In several studies it has been estimated that about 10 to 15 percent of the youth between 15 and 25 are cultural drop-outs (BMJFG 1981, p20)***. The number of people who sympathize with the main normative syndromes of this core group is much higher of course, as can be seen from the attendance at certain demonstrations, for example. The importance of this movement stems from its public significance and political potential which come near to something like a veto of official political decisions—desite the fact that it comprises only a minority of citizens.

(ii) The new cultural movement is being politicized in some places already, and then turns to violent and "non-conventional" activities. This is not only the case with demonstrations, but also with "ambiguous" parties and parliamentary representatives†. "Youth is becoming more and more brutal" is the simple conclusion of a recent review of empirical investigations (Ernst 1978, p28).

---

*Recent developments in the UK have shown that this perspective is not only a German one. See e.g., *International Herald Tribune* (Paris) 7 July 1981, p1 about the Liverpool riots. These were commented on by a German journalist in the following way: "Liverpool—there is a far-reaching agreement upon this—was no racial confrontation in the narrow sense. It was the remonstration, animated by gloomy anarchism, of an alienated, hopeless, bored youth living in strained relations with their parents as well as every authority, without any direction, and regarding every orderly organized society as an enemy" (T. Bode, Aufbegehren einer entfremdeten Jugend, in: *Süddeutsche Zeitung*, 8 July 1981, p4; my translation).

**Furthermore: "More and more young people no longer understand and accept the traditional understanding of politics. They are indifferent to the parties; the "grand politics" is believed to be too far away from the immediate sphere of living that they can grasp, and cannot directly be influenced therefore. Some of the grand political issues, such as the movement for peace are highly attractive with respect to their moral implications; yet they become relevant to activities apart from demonstrations only if they can be brought within the sphere of personal experience" (T. Bode, Aufbegehren einer entfremden Jugend, in: *Süddeutsche Zeitung* 8 July 1981, p4.

***In the last communal elections held in the city of Frankfurt in 1981 the proportion of voters below the age of 25 who voted for an ecologist party was 25 percent (*Frankfurter Rundschau* 15 May 1981, p13).

†A new parliament was elected in 1981 in West Berlin which bears the status of a state ("Land") within the FRG. In this election the so-called "alternative movement"

(iii) Another German (only?) development has to be mentioned, which does not appear as revolutionary as the "alternative culture" movement, but which is at least as influential: political decisions are more and more frequently being over-ridden by court decisions*. "Juridification of politics" could be a suitable name for this phenomenon, and it is particularly alive in, yet not restricted to, the field of nuclear power, a field that risk research is traditionally very much concerned with. People who provoke these court decisions (which are binding on political institutions according to the FRG system of administrative law) thus demonstrate (a) their "alternative political consciousness", and (b) their (often desparate) unwillingness to use ordinary political channels for their exercise of influence**.

9. RISK RESEARCH IN A SOCIETY WITH WANING CONSENSUS

Risk research, as an applied research effort, was invented to (and attempts to) support political institutions in their job of mediating technological projects with social needs and (deplorably dominating***) social opposition of acceptance, respectively. Figure 1 shows the position of risk research in relation to the political system and society. It additionally clarifies the focus of this paper as compared with the amplification interests which can be found in risk research debates at present. In view of this functional status of risk research, it is obvious that the unstable position of political institutions within society must also affect risk research. If my hypothesis is correct, that the political institutions are becoming less and less able to cope not only with risk distribution, but also with society as a whole, then risk research has to suffer from this development too. So, risk research is bound to a political game that is fundamentally opposed as such by the new social movements. Two conclusions can be drawn from this.

---

succeeded in gaining parliamentary seats for the first time (there are already ecology representatives in two other FRG states, Baden Württemberg and Bremen). This entrance into parliament was near to a revolution: the "alternatives" got 7.2 percent of the votes, considerably more than the proportion of the smallest (third) established party in Germany, the FDP, which got 5.6 percent. The highest proportion that the "alternatives" reached in a single electoral district (Kreuzberg) was 18.2 percent. In Berlin there was a merger of the "alternatives" and the ecology movements. "Alternative" orientation means a certain kind of radical opposition to representative government and parliamentarism. Those representatives who are now members of parliament regard themselves as part of a grass-roots movement. Consequently, a free mandate is not given to the "alternative" representatives; instead they are bound to decisions of their "basis". The actions of this "basis" are often characterized by force (cf. H. Schuster, Parlamentarische Verteidiger der Gewalt, in: *Süddeutsche Zeitung* 30 July 1981, p4). An analysis of the Berlin elections came to the conclusion that the increase in the "alternatives" can be interpreted as a tendency to build up something like an alternative "milieu" which may lead to a functional confrontation with established parties in the future (cf. Forschungsgruppe Wahlen, Wahl in Berlin, Mannheim, 14 May 1981).

*H. Riehl-Heyse, Mutmassungen über die Wirbelschleppen, in *Süddeutsche Zeitung* 2-3 May 1981, p10.

**"It is after all due to a change of opinions and the loss of consensus in basic questions, particularly in the field of environmental policy, that parties and interest groups, exasperated citizens, and also grumblers, with increasing frequency appeal to the courts as arbitrators today" (H. Schuster, Versuchung des Rechtswegestaates, in: *Süddeutsche Zeitung* 27 April 1981, p4; my translation).

***It is worth mentioning that in business administration theories the term "risk" is taken as a neutral category: "Although in modern parlance the word risk is usually associated with "hazard" or "danger of loss", economists use the term to denote chance or luck, *both* good and bad" (Sarnat and Engelhardt 1981, p107).

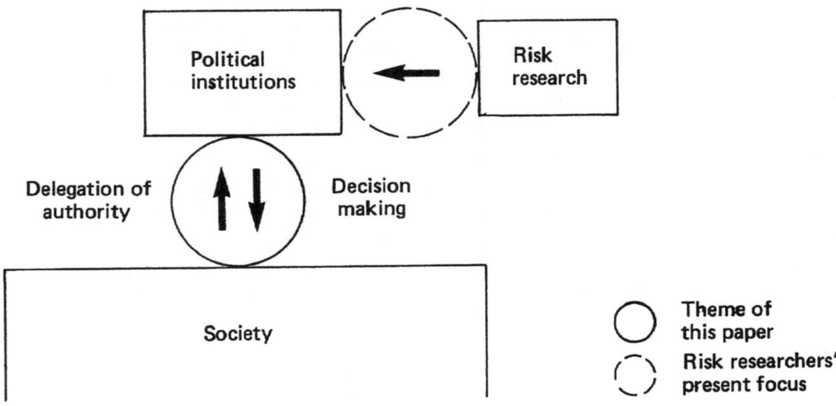

Figure 1.

(i) Risk research as an effort of research applied to political (decision making) institutions will not be capable of changing a social reality that turns against traditional political institutions, however valid its scientific knowledge might be. It does not make up more than a small piece of advice in policy making, i.e., it falls into the input phase of decision making, and is thus a victim of weak political institutions.

(ii) Risk research is suffering from the fact that, as a mere applied science, it is missing some basic research that would help to loosen its ambivalent connection with political institutions and could lead to more independent, "open", and "societal" paradigms of risk. Let me explain this idea further. Cost-benefit analysis, the field of risk research, without any necessity of risk research, has been over-ridden by real social developments: there is a tendency of social *groups* to be divided along the cost and benefit lines; the academic calculus has turned into a real social clash and gap. In other words, the cost-benefit calculus is being made up more and more by social groups promoting clear-cut interests, pros and cons.

10. RISK RESEARCH ALTERNATIVES

Two alternatives are open to future risk research in my view.

(i) Risk research transcends its narrow political insitutional perspective and begins to integrate risk perceptions and interests important in society—even if these are opposed to the official perspectives of political institutions.

(ii) Risk research transcends its narrow technological projects paradigm and begins to integrate concepts such as the one that regards political institutions as the main risk to society today.

REFERENCES

Baum, G. (1981) Rede zur Eröffnung der Arbeitstagung "Technologische Risiken und gesellschaftliche Konflikte. Bonn.

BMJFG (Bundesminister für Jugend, Familie und Gesundheit)(1981) Zur alternativen Kultur in der Bundesrepublik Deutschland. Bonn.

Conrad, J., and Krebsbach-Gnath, C. (1980) Technologische Risiken gesellschaftliche Konflikt. Politische Risikostrategien im Bereich der Kernenergie-Materialband. Frankfurt: Battelle-Institut.

Ernst, T. (1978) Daten zum politischen Extremismus, in: L. Unterseher (Ed.), Junge Leute von Heute (transfer 5) Opladen, pp28-35.

Heller, E.D. (1980) Umfragen zur Kernenergie. Analysen demoskopischer Ergebnisse. Frankfurt: Battelle-Institut.

Hovden, J. (1980) Accident Risks in Norway. Oslo: Risk Research Committee, Royal Norwegian Council for Scientific and Industrial Research.

Inglehart, R. (1977) The Silent Revolution. Princeton, New Jersey.

Inglehart, R. (1979) Wertwandel in den westlichen Gesellschaften: Politische Konsequenzen von materialistischen und postmaterialistischen Prioritäten, in: H. Klages and P. Kmieciak (Hrsg.), Wertwandel und gesellschaftlicher Wandel. Frankfurt, pp279-316.

Kaase, M. (1976) Bedingungen unkonventionellen politischen Verhaltens in der Bundesrepublik Deutschland, in: P. Graf Kielmansegg (Hrsg.), Legitimationsprobleme politischer Systeme (Politische Vierteljahresschrift, Sonderheft 7/1976). Opladen pp179-216.

von Krockow, C.G. (1979) Ethik und Demokratie, in: Aus Politik und Zeitgeschichte. Beilage zur Wochenzietung das Parlament 49:3-21.

Lathrop, J.W. (1981) Evaluating Technological Risk: Prescriptive and Descriptive Perspectives. This volume.

Mandl, C., and Lathrop, J. (1981) Assessment and Comparison of Liquefied Energy Gas Terminal Risk. WP-81-00. Laxenburg, Austria: International Institute for Applied Systems Analysis.

Raschke, J. (1980) Politik und Wertwandel in den westlikchen Demokratien, in: Aus Politik und Zeitgeschichte. Beilage zur Wochenzeitung das Parlament 36:23-45.

Ronge, V. (1980) Theoretical concepts of political decision-making processes, in: J. Conrad (Ed.), Society, Technology, and Risk Assessment. London, pp209-238.

Sarnat, M., and Engelhardt, A. (1981) The pattern of risk and return in Germany, selected industries, 1928-1976. Konjunkturpolitik pp105-125.

Schindel, V. (1978) Risikoanalyse. München.

Schissler, H. (1979) Der Werteaspekt in der Forschung zur politischen Kultur. Ansätze und Ergebnisse, in: H. Klages and P. Kmieciak (Hrsg.), Wertwandel und gesellschaftlicher Wandel. Frankfurt, pp351-364.

Sontheimer, K. (1970) Kompromiss, in: Hanbuch des deutschen Parlamentarismus. München, pp245-247.

Institutional Mythologies and Dual Societies in the Management of Risk

Brian Wynne

University of Lancaster, Bailrigg, Lancaster LA1 4YN, UK

1. INTRODUCTION

The risk assessment fraternity is well aware by now of the limited role played by various "technical", "rational", or "analytical" approaches to risks in real decision making contexts*. Reactions to this news vary from the familiar technocratic lament about the irrational ways of the world outside their rational bastions, to the vigorous celebration of the political and psychological "underworlds" which impose those limits on the technocratic perspective. Eschewing these polarized extremes, however, most people in the field are still working out their responses, as are those like myself who are really marginal to it, and who came with an already well developed experience of the sociological dimensions and instrumental limitations of scientific rationality. Given that reactions are still in the formulative stage; given the centrality of the idea of scientific rationality in risk assessment philosophies and methodologies (and in modern policy making generally); and given that appropriate responses are of practical importance in policy making; I shall attempt in this paper to outline some reasons why the sociological analysis of scientific rationality that has developed to some maturity in the last few years is directly relevant to the kinds of questions now facing risk assessment when it attempts to define where it should be heading next. In the process I shall outline some areas of research under the general area of risk assessment of technology politics that this different perspective suggests. There is a potentially important though as yet unrealized correspondence between many of the insights from the psychological research which has influenced risk analysis and the sociological analysis of scientific rationality. In presenting the latter my aim is not to upstage the former but to point out how the naturally individualistic learnings of psychology can and must be developed into sociological and anthropological frameworks for formulating questions and possibilities. If there is one central concern of this paper it is the following.

Several authors have by now recognized the "pitfalls of analysis" (see Majone and Quade 1980), whereby analysts tend to see what their tacit assumptions and values predispose them to see, and whereby "heuristics" introduce "biases" in their work (Tversky and Kahnemann 1974). It is of the utmost importance that these are recognized as part and parcel of science, not as eradicable lapses from proper rational scientific procedures. There is a pervasive myth about the nature of science which supports this false approach to the question of "analytical bias." The tendency in the literature is to regard bias or mistakes as individual and isolated in origin, which suggests that ideal objective scientific knowledge can be attained in professional practice and as an input to policy issues. As I shall argue, this gives a fundamentally misleading and politically damaging picture of the role of expertise, and may make us part of the problems we analyse.

There can be no doubt about the central role played by science in risk assessment. Even when political factors have been recognized as relevant and legitimate the attempt has always been made to define a clear boundary between the two realms,

---

*See, for example, the range of views in Conrad (1980), Otway and von Winterfeldt (1980) and Lathrop (1980).

such that scientific analyses of "real" risks and comparative risks could be an "independent", objective input to the political process. However, although there might be agreement that the role of science has been central, there would, I suggest, be much less agreement as to what that role is, and indeed, how many roles science may be playing at the same time in the processes of political communication. Thus for example there is the obvious methodological role in systematic observation, quantification, and calculation. This is largely instrumental in that it is employed deliberately and directly to achieve explicitly defined objectives. However there are less obvious processes going on when such scientific methodologies are used and displayed, and as is often the case, the less obvious these processes are the more important they may be. Some of these less explicit roles are supported by generally established suppositions as to the nature of science—the kind of authority it can naturally provide—which need to be dug out and dispassionately examined. There are two broad conclusions towards which I would like to direct this discussion.

(i) There is a tacit symbolic role played by images of scientific rationality and control projected in the risk assessment-policy process; this is at least as important as its purely instrumental-analytical role. In other words, decision making elites are creating and elaborating myths of order and authority based in "science" that are essentially therapeutic and legitimatory. These myths about the nature of science and thus about the power and scope of analysis tend to justify the dominant role of professional elites whilst encouraging reassurance and quiescence amongst the public at large: the former are not deliberately foisting those myths upon the latter—they are themselves enmeshed in those myths.

(ii) One general consequence of the false image of science portrayed in this kind of symbolic action is the tendency to see errors or mistakes in scientific or technological systems—e.g., accidents—as sporadic individual lapses from normal standards of objectivity and rationality, essentially eradicable. This basis of symbolic reassurance at the political level reflects a direct analogy with the rationalist model of science in which the only alternative to complete objectivity is individual subjectivism. Thus departures from an idealized absolute scientific objectivity—be they accidents or dogmatic outbursts from scientists advocating a particular policy—are taken as sporadic lapses from proper standards of conduct, not as *systematically* generated in the very structure of organized scientific thought or technology. The associated risks can thus be defined as essentially isolatable, and the basic approach can never be falsified since all like instances can always be defined in the same sporadic terms. The sociology of science as developed in Europe since the early 1970s has challenged this rationalist myth of science and in doing so has offered insights of direct value to fields such as risk assessment which are practised in the midst of these deep symbolic political transactions about science, rationality, and order*.

---

*A concise summary of this field, and a discussion of its departures from previously dominant, rationalist models of science, is Mulkay (1980). This tradition of research is best illustrated in sustained fashion in the journal *Social Studies of Science* (Sage). See especially the special issues, Vol. 6, Autumn 1976, and Vol. 11, February 1981. Also highly relevant are Latour and Woolgar (1980) and Barnes and Shapin (1979). An author sensitive to these dimensions of science who has applied them in some ways to politics of science is Ezrahi (1974); see also Ravetz (1974). For a thorough philosophical analysis of science from this general perspective, see Hesse (1974).

2. THE MYTH OF NATURAL CONSENSUS

A central element of the social image of science has been the idea that science reaches a natural consensus by following definite, universal principles of method, logic, and judgment. Any conflict is thus taken to be the temporary result of illegitmate "pollutions" of one kind of another—ignorance, incompetence, prejudice, ideological motivation, etc. There is one sole objective truth dictated by reality, and all else is pathological deviation from it. Science inexorably purges these passing deviations by continual critical self-correction, so this view goes. It merely discovers the meanings already inherent in the facts that are waiting to be laid bare. This image of science has dominated public and scientific discourse, and as with most aspects of domination, it has done so by being taken for granted. It has been entrenched in the roots of population and political culture by scientists and social analysts of science (Cameron and Edge 1977; see also Mulkay 1980 and Ezrahi 1974).

Technology assessment and systems analysis lived (and have arguably died) under the spell of this myth, and one can see how the same currents of thought exist—albeit in tandem with a healthy air of scepticism now—in the risk assessment field*. As in all systems of myth, the response to evidence which might contradict the basic assumption instead leads to an intensification of that founding faith by further elaboration of it. Thus, for example, the implicit justification of the sophisticated psychological research on attitudes to risks has been to defend the quantitative risk-comparison/revealed-preference approach from the evidence that it does not accord with reality, by defining those who persist in evading such quantitative, externally imposed frameworks of meaning as chronically ignorant or malign**. In other words, the reaction to conflict at the first level of analysis has been to assume that some of those involved have not "seen it correctly", and that educated by further research they would do so. That they have not done so has not led to the accepted falsification of the myth of natural consensus, but to its escalation in the intensified search for more and more subtle psychological factors to "explain" the "irrationality" of resistance. Although it would be unwarranted to attribute this assumption to all the researchers in this field, it is strongly implied in much discussion and use of the research***. In principle this myth could be elaborated without end if it plays a sufficiently important tacit role in supporting the status and power of important interests, as appears to be the case.

One can see the parallels of this mythology in the general field of technology policy making. In her pioneering study of the Cayuga Lake nuclear controversy, Nelkin described how, contrary to the expectations of those concerned, the conflict was not resolved by allowing the opposed experts to debate with one another outside the hysteria of the public media (Nelkin 1971). Instead this only produced more technical elaboration and polarization. Based upon the myth outlined above, the predominant reaction has been to assume that the "real" natural consensus that should have emerged had been blocked by the intrusion of illegitimate, extraneous pollutions, such as the media, pressure groups, etc. Yet when these were removed, no such purification and consensus took place. The further reaction to this first-order process has been to assume that one side or the other of the experts themselves has introduced the pollution of proper science. Thus allegations of lack of

---

*For an early critique of technology assessment from this angle, see Wynne (1975) and Winner (1977); on systems analysis, for example, see Hoos (1969), Boguslaw (1972) and Weizenbaum (1980).

**Thus for example, the perspective is repeated endlessly in the UK nuclear industry's mouthpiece *Atom*. See, for example, the exchanges between D. Pooly of the UKAEA and C.H. Green in *Atom* vol. 295, May 1981.

***See Starr and Whipple (1980) and Okrent (1980). See also the proceedings of the 1980 conference "The Assessment and Perception of Risk" (to be published by the Royal Society of London). Starr and Whipple, for example, see *political* conflict only in the mismatch between public benefits and individual risks, with apparently no possibility of more basic conflicts about the "public good" between different publics.

proper qualifications, downright incompetence, ignorance, or ideological motivation have been the main currency of debate, not the facts themselves. This has been repeated over and over again in expert disputes on such issues (Doty 1972, Robbins and Johnson 1976). Indeed legal cross-examination in the increasing number of forums which adopt the legal framework as a method of resolution in these matters enshrines the belief in the justification for (scientific) amateur cross-examination of experts. The general assertion that policy conflicts are due to residual—if often chronic—technical uncertainties is a further manifestation of the myth in that it implies that the full scientific facts would resolve the conflict one way or the other. On this view it is not conceivable that the scientific positions could themselves be permeable to different social perspectives that cause the associated scientific positions to remain incompatible however well developed the field might become.

In the terms of Figure 1, the absolute boundary between science and politics or fact and value is taken for granted as a metaphysical principle*. In the terms of the basic myth about science that I am outlining, if scientists begin with a variety of different views, A, B, and C, then a properly conducted rational debate will lead them to a single truth, T, which can then be fed into the (supposedly entirely separate) political system, for a possible variety of different policy conclusions, W, X, Y, Z, according to the democratic expression of social values. When T remains obstinately obscure, the first reaction in the natural consensus myth is to allow the experts to debate away from the pressures and temptations to play to the crowd, and public media; when this fails, the next step is to try to expose which side of the expert dispute is incompetent, malign, etc.; when this fails by scientific debate, the legal process often takes on the explicit belief that the scientists have failed to provide consensus either because the political undercurrents run even deeper into the views of one or the other group of experts than previously imagined, and need the ultimate test—legal cross-examination—to expose them, or because scientists are not clear in their expression of the truth and need the unsurpassed precision of the legal framework to *clarify* for them where the truth lies. That there exists a single truth beneath these several layers of extraneous camouflage is never doubted, at least in the rhetoric of what is being conducted and why. In other words, so the myth goes, if we employ rigorous enough analytical weapons, we will eventually purify the murky waters where science and politics are dissolved into one another, and define the clear-cut interface properly**.

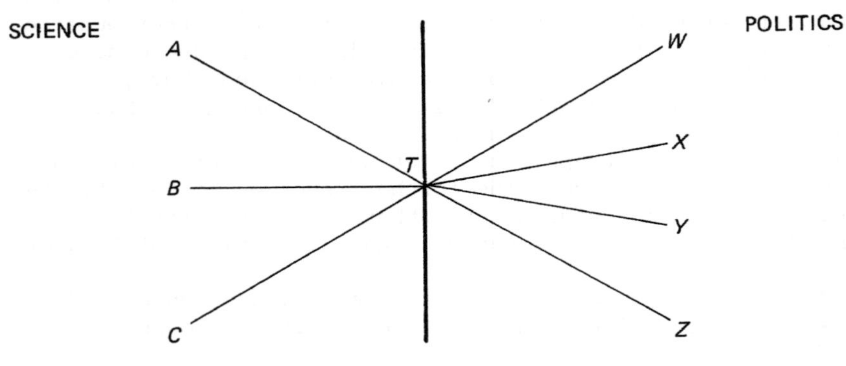

Figure 1.

*Even the Kemeny Commission, so refreshingly alert to the persistence of limiting "mindsets" in technological organizations, relayed the taken for granted idea that technical questions could be resolved in isolation from politics, then fed into the categorically separate "political hopper" (see Kemeny 1975).

**I have documented this in Wynne (1981a). Assumptions about the value of a legal framework are embodied in the science courts proposal mooted in the USA. For a review see Mazur (1977).

This mythology has of course been reassuring to professional experts concerned to avoid the ambiguities involved in playing a political role. It has allowed a constant stream of such groups to define such a role artificially as unpolitical, and thus noncontroversial. It has been equally valuable to their political patrons in the power structure, by defining conflict as pathological deviation from a singular, objective order based in a truth which "properly controlled" experts can discover (Habermas 1975). Under this framework, legitimate political conflict over fundamental social aspirations becomes difficult even to envisage, let alone sustain, without recourse to entirely alien languages such as withdrawal or violence.

3. THE MYTH OF SPORADIC ERROR

As the argument of the previous section made clear, a central element in the dominant mythology of rationality is the belief that conflict originates in departure from rational judgment or conduct. We can draw a direct parallel here with a dominant attitude to risks arising from complex technologies. After an accident there is always a process of interpretation to establish causes and responsibilities and to learn lessons for the future. In the cases of two accidents in the nuclear industry, even though they are separated by 22 years in time, one can see the suggestion of an interesting pattern which throws up some important questions for policy research. The Windscale fire of 1957, which released over 20 000 curies of airborne radioactivity, began in the graphite moderator of the air-cooled nuclear pile, when the Wigner energy release was allowed to get out of control. The Three Mile Island accident in 1979 released a few tens of curies of radioactivity, but in these less deferential times caused orders of magnitude more political unrest. In each case a critical stage in the onset of the accident was the organizational neglect of important operating knowledge. In each case a central part of the *post hoc* interpretation was the assertion that the operators had acted in an incompetent, inadequately trained fashion. Again in both cases, an obvious response was therefore to improve recruitment, training, and general standards in the operational fields concerned. Thus there is made to appear a clear-cut problem with a clear-cut solution. Everyone can reassure himself that the lack of control was an arbitrary, individual lapse from proper standards, but that such occasional occurrences can be fenced inside an ever-tightening circle and isolated by better training and (in the TMI case at least) better psychological selection and conditions*. Yet a different interpretation of each accident is available. In the Windscale case the interesting question is whether the operators and their superiors knew about the Wigner release and its significance, and how to deal with it. If they did, but just had a bad day, then a picture of control could be portrayed, and the mistake seen as sporadic, not systematic. On the other hand, if they did not, then a more difficult problem of public explanation and justification presents itself. For in such a case either the organization as a whole was operating in ignorance of a central piece of technical knowledge or "it" knew, but failed to convert this into practical operating knowledge. Either way, public reassurance about an industry based for its prestige upon an image of scientific rigor, control, and authority, would be difficult to retrieve. Not suprisingly perhaps, *post hoc* accounts differ sharply on this very point. Some accounts said that the operators made a mistake in what should have been a routine operation. The blame could thus be clearly laid and exorcized in the assurance of better training and recruitment. However when it is learned from more detailed scrutiny that their mistake was encouraged by the lack of an operating manual and by various design blunders such as control thermocouples not being placed at the hottest part of the core, and the graphite blocks being spaced too far apart to dissipate excess heat by conduction, the origins of the accident appear

*On the Windscale fire, see Welsh and Wynne (1981). On Three Mile Island, see Kemeny (1979), Nelkin (1981); *Journal of the Electric Power Research Institute* special issue on TMI, June 1980. An indication of how shallow-rooted the "mindset" problem was conceived to be by the industry was that it thought it had already solved the problem before the Kemeny Commission reported. See Zebroski (1980).

to be more complex, and indeed fundamental to the system (for documentation, see Welsh and Wynne 1981).

A very similar analysis can be made of the TMI fiasco. Here, the sequence built up due to several initially small errors. A key mistake was the decision by the operators to turn off the emergency core cooling system when the pressurizer water level was rising, even though core pressure was falling. They did not know that the relief valve that had opened to release pressure, when a pump failed elsewhere in the circuit, had stayed open. Their instruments said that the signal had been relayed to re-close it. Thus the reactor core was losing its water coolant whilst they had switched off the emergency cooling water supply. On the face of it they acted incompetently, and later comment repeatedly asserted or implied this*. In the post-accident period the two reactions that have overwhelmed all others have been the fervent pursuit of better psychological control of operator reaction to boredom, stress, and operating room instruments, and better recruitment and training. Again, however, this neglects some important factors. Prime amonst these is that until a short time before TMI, no one had ever foreseen the possible combination of events, especially the anomalous apparently mutually contradictory indications from the core, of rising pressurizer water levels and falling pressure. When this was seen at the Davis Besse plant 14 months before TMI, and when an engineer asked that the possibility be widely advertized and incorporated into standard operating knowledge, the information was dispatched into oblivion within the organization, even though the key event, the relief valve's sticking open, was apparently a relatively common occurrence.

Thus there are two main aspects of the TMI "explanation". First, the operators are said to have acted incompetently, encouraged by bad control room design. They lapsed from "proper" rational norms. Better training and recruitment and better psychological "fixing" was claimed to clear up this problem. The implication is that the scientific knowledge for control *was* available as an objective body of knowledge, but that the operators failed to enact it. Control was there in principle, just lost by an individual, isolated error. This entirely conceals the fact, as in the Windscale fire of 1957, that the scientific knowledge, the basis of supposed control (and public reassurance), was *not* available. Second, the organizational neglect is itself virtually neglected or treated as an instance of "complacency" to be eradicated by more "vigilance" and associated moral disicpline. Yet these both conceal critical points about control and its possibilities. The first conceals the fact that within the paradigm of existing understanding, the operators acted in exemplary rational fashion. What was significant was not their alleged deviation from rationality or competence, but the limited scope of the prevailing knowledge—a much more general and fundamental problem, and one (like the Windscale example too) which, if exposed, would undermine the covert symbolic, legitimatory role of the public image of a technology controlled by the wisdom, discipline, and analytical power of science. Likewise the idea is concealed that a technological organization may not be able to select, convey, and transform significant information into practice without regularly making classification mistakes between the trivial and the important. That scientific expertise generally does not itself hold any formula for doing this is an insight into the limits of scientific knowledge and rationality that if made available in public language would undermine the scientific myths upon which such commitments are legitimated. There is no reason of course why the wider publication of this idea of science as incorporated in technological organizations should *necessarily* engender a loss of legitimacy, and thus opposition, except that the public has been encouraged to assimilate those myths for some considerable time now, and amidst a pervasive network of commitments. In any case what is significant is that the positivist myth of scientific rationality has encouraged the assumption that really there are no such things as "organizational risks" systematically rooted in institutionalized faith, arrogance and dogma, only sporadic risks from badly adjusted or over-provoked individuals (or from malign elites, a version of the same scientistic myth). This profoundly unsociological

***The Journal of the Electric Power Research Institute* special issue on TMI, June 1980, exemplified this attitude most strongly.

perspective is consolidated by the faith that conflict and "deviation" arise in rational institutions only when someone acts pathologically*. There is supposed to exist a single objective route marked "truth" or "control" and sporadic deviations can be clearly demarcated, explained, and contained.

4. THE MYTH OF INSTRUMENTAL PURITY

An alternative way of phrasing the familiar idea that science can be clearly demarcated from politics is that facts can be categorically distinguished from emotions or values. Facts lie in the objective realm of rigid logic and empirical observation; emotions and values lie in the subjective realm of nonempirical feeling. In the first category, meanings are taken to be intrinsically given—dictated by nature; in the second, meanings are allowed to be invested in things by active personal feeling and social influence. Thus a categorical distinction has come to be taken for granted in this perspective, between instrumental calculative forms of knowledge such as law and science, which have trustworthy objective authority, and symbolic forms of knowledge such as poetry, art, and religion, which are suspect. Scientific knowledge, so this view goes, is not constructed of symbols whose meaning might be socially malleable, because it is fashioned from the cold steel of hard empirical fact**.

One of the insights which recent sociology of science has begun to develop to some maturity, however, is that even scientific knowledge in disciplines that appear to have little to do with social affairs is pervaded with meanings and tacit explanatory purposes that go far beyond any immediate purely scientific, analytical or instrumental concerns. MacKenzie (1981), for example, has shown how modern statistics was developed with strong tacit interests in eugenics in the minds of its pioneers. Work on modern quantum theory has argued forcefully that its pioneers in Weimar Germany were influenced by the tacit need to protect a radically new, anti-mechanist, anti-determinist public image in a vigorously anti-scientific cultural climate (Forman 1971). This engagement in symbolic language may therefore have encouraged them to think in the anti-causalist terms that the new physics reflected. Such examples could be repeated many times from recent sociology and history of science (Mulkay 1980 gives the best summary review).

Thus although the essence of science is regarded as the methodical purification of all extraneous, nonempirical factors from the body of knowledge, we find that even the exemplars of pure science are imbued with wider symbolic currents as well as instrumental-analytical power. When one moves to scientific fields involved in political issues, such as risk assessment, these symbolic dimensions cannot but be even stronger. Thus for example the debates over the safety of many environmental pollutants such as radioactivity, lead, or 245-T, are structurally conditioned by the fact that underlying the overt technical discourse is a symbolic discourse in which those who have previously committed themselves to a particular scientific point of view, with particular policy implications, are attempting to defend their long-term credibility, and the credibility and public authority of their decision making institutions. This kind of tacit symbolic structuring of the issue may be of greater practical importance than any purely technical-instrumental discourse since at this latter level there are always likely to be more ambiguities and diverse

---

*The analogy with the positivist philosophy of science is seen most clearly in Scheffler (1967) and Mitroff (1974). Mitroff's sociological study reflects the typical early tradition in especially American sociology of science, which thought that the scientific truth comes "naturally" to scientists, so that the only thing to explain was error. For the sociological inspiration of this general perspective see Merton (1975). For a specific critique, see Barnes (1972).

**This assumption that symbolic knowledge is entirely unempirical, whereas instrumental knowledge is empirical has pervaded sociological and anthropological thought too. See Gellner (1974) and Firth (1973). For a critical view more closely aligned to that of this paper, see Sperber (1975). The work of Mary Douglas (1970, 1975) has been a central point of reference in this area.

interpretive options available to the expert participants than the rationalist idea of science can recognize. If risk assessment processes are unable to identify these symbolic levels of political transaction with respect to scientific expertise, they are likely to be irrelevant from the outset. Again, another general area of research suggests itself, once the myths in which risk assessment itself is entangled are cleared away so that this symbolic undergrowth is uncovered. The point which cannot be strongly emphasized is that social analysis of science indicates that there is no warrant for the general convention that instrumental-calculative and symbolic dimensions of knowledge can be disentangled. We must also therefore discard the assumption that the latter is an inferior language which conveys nothing of authentic *social* meaning, only individualized subjective meanings, and that it arises only as a nonempirical pollutant of true, authentic public knowledge.

5. THE MYTH OF SCIENTIFIC REALISM

The rationlist account of science has naturally tended to consolidate the belief that science comprehends reality and that all other forms of knowledge are irrelevant to issues of public policy. It is a matter of historical fact that the economic or military pay-off in some fields of academic science such as nuclear physics, electronics, certain chemical areas, etc., allied with the unprecedented prestige granted to pure science in the post-war period, meant that academic research came to be established at the most influential points of technological commitment and policy decision making. Niblett (1979) has shown how various advanced industrial companies in Britain used direct links with prestigious academic research institutions not primarily as an instrumental arrangement for better exploitation of R and D, but as a symbolic weapon to advance their own prestige and autonomy by association with potent general symbols of authority and legitimacy, namely academic science. This further example of the symbolic transactions underlying the explicit language can be applied to the nuclear case again. One of the recurrent themes in otherwise quite different analyses of the British nuclear program has been the monopoly role of the UKAEA, and the domination of policy choices by the academic scientists who have been in charge at Harwell (Williams 1980, Burns 1979). The only successful period of the British program is often treated nostalgically as a "golden age", when the brass-tacks engineer Hinton was in charge of production in the 1950s. It is plausible to interpret one of the central failings of the British nuclear program as the naive lack of realism of the abstract, theoretical design-oriented scientists when it came to scaling up parameters and implementing theoretical constructs and conditions in actual engineering and commerical practice, and (increasingly significant) in organizational and social practice. This might not have been such an important problem had they not dominated the policy process, but the point is that they did so dominate it and were so divorced from these more mundane realities because of the symbolic, legitimating role that their prestigious pure science image was playing in the public domain. In this case therefore, one can see how the process of technological commitment and innovation was established in a socially rarefied realm where most of the relevant questions to do with those commitments were not conceivable, let alone answerable, because the politically insulated, academic science culture was not capable of thinking in such terms. Although the nuclear example may be an extreme one, it does not seem to be atypical of the general dislocation between the point of innovation and commitment where conceptually narrow scientific world views prevail, and the points of social enactment where the exposure of the inadequacies of those world views comes too late and is too disorganized to be constructive or influential. The state of innovation is protected from wider participation by straightforward secrecy and proprietorial control of scientific research, or by ideologies about the power and unpenetrability of science. These are bolstered by the myths about rationality outlined here.

An example of the same insulation of technological commitment from social realities that should be regarded as part of the commitment question can be taken from the issue of the safety of 245-T. In the face of growing evidence that the spray induces cancers in those using it, the British Advisory Committee on the

Safety of Pesticides asserted that it was safe, assuming that the proper instructions for production and use were followed correctly. Yet, as the National Union of Agricultural and Allied Workers pointed out, many workers were told to use the sprays with no instructions, and no protective gear. Their theoretical right to demand these would result in harrassment and dismissal on many farms. The theoretical assertion by the expert committee, of the practical safety of 245-T (and even this theoretical assertion is contested), was thus based upon utterly unrealistic assumptions about 245-T *as socially enacted**. The committee refused even to hear evidence from the union on this matter, presumably because to have done sp would have been symbolically to confer legitimate status upon this general category of evidence in principle. This would be to undermine the monopoly role of their committee and its definition of science in defining the risks and even the relevant questions. Their position was bolstered by the myth that science merely uncovers reality as it objectively exists, and does not artificially reduce issues down to conceptually narrow and reduced problems to which answers can be given. If there is no recognition that anything is left out of account in such conceptual structures, there can never even arise the question of whether what has been left out is more important than what remains. There are many parallel cases to that of 245-T that call for research in terms similar to those outlined here.

6. THE MYTH OF ACTIVE CONSENSUS

Another key component of the rationalist model of science is the idea that scientific consensus is generated and continually self-corrected because scientists engage in unremitting "organized scepticism"—every proposition is critically scrutinized and vigorously tested to attempted falsification before it is granted the status of accredited knowledge (see, e.g., Merton 1975 and the critiques referred to by Burnes 1972 and Mulkay 1980). Consensus is therefore the product of *active* critical espousal of beliefs. This image of scientific consensus has been a powerful if largely tacit influence upon conceptions of political consensus. Yet the sociology of science has shown that the typical mode of scientific consensus is one in which scientists accept on trust most of what they believe as true, via their own established channels of social authority. Consensus is achieved, as much if not more, by default of criticism rather than by active testing and espousal. This immediately implies a less authoritative kind of "consensus" and entirely different perspectives upon political processes.

One political analyst who has recognized and investigated this alternative way of looking at political consensus is Edelman (1971)**. He has argued that political

*I have documented this from the national press. It is a case study overdue for research in the broad terms I am suggesting.
**See Edelman (1966, 1971, 1977). For all the genuine value of the anthropological perspective offered by Edelman, the pervasiveness of the positivist myth of rationality is illustrated by the fact that even he falls foul of it. Thus there is a tendency to assume that whereas ordinary people live in a manipulated world of symbols that have no link with empirical social reality, elites are busy doing the manipulating with complete instrumental clarity, calculating their own interests and the means of securing them. Thus there is a suggestion that pure rationality can and does find concrete attainment at that rarefied level. The fact that Edelman does not like it is not relevant. The point is that he appears not to see that the elites too become bound up in their own myths. Winner, for example, describes very persuasively how modern society has woven the myth that technology is autonomous from social forces, so as to rationalize an unwillingness to exercise responsibility. Yet this widespread deterministic outlook finds expression alongside its opposite—the idea that modern science has provided historically unique opportunities for actually controlling our destiny, and for transcending political ideologies. It should not be surprising if we were to find that the anti-determinist attitude were being elaborated into a self-justifying mythology among decision making elites, whilst this deterministic perspective was being articulated among "the masses" to

elites regularly engage in the manipulation of the symbols of public language to persuade people at large that their environment is being protected; that industry is being properly regulated; that key cultural norms are being enforced; when in reality these may not be being fulfilled. Edelman analyzes this as a process of symbolic action to reassure people of what they want to hear, that life is secure, and that they can sink back into a state of political quiescence and overt apathy, even when they are being exploited under this rhetoric of "protection", etc. Far from expressing their wants forthrightly and critically examining whether or not they are being properly pursued, most people are typically confused, open-minded, and inconsistent about their wants, and have only limited interest in pursuing them steadfastly. Edelman distinguishes between two fundamentally different analytical approaches to political processes. On the one (dominant) hand, the positivist approach takes it for granted that politics is the aggregation and realization of the wants of many diverse people, whose values and wants are relatively clear and stable, and who examine and judge political policies and leaders according to who best fulfils them overall. Policy analysis then becomes a case of identifying how best those wants can be aggregated and realized. In our field, we insert the extra stage, of the education of those wants and processes via research into the implications of alternative choices; however this provides the ambivalent link with the radically different, anthro-political approach advocated by Edelman. In this perspective people's wants are recognized as less formalizable and concrete, and political utterances, policies and processes themselves are seen to cue these "wants" into certain forms that are, at best, passively tolerated. These political events are themselves to be seen as the paraphernalia of symbolic action that encourages quiescence and consensus-by-default, but which conceals this in the explicit language of the positivist image of consensus drawn from "activist" models of scientific consensus.

This distinction is fundamental, and directly relevant to risk assessment research. The pedigree of risk assessment lies entirely within the positivist tradition. In this it is in good company, since the same can be said of nearly all that passes for policy analysis, and indeed most political research generally. By definition, the risk-benefit framework of analysis involves the assumption that benefits and costs are definable against a positive yardstick of concrete social wants. The use of the social cost-benefit framework in attempting to develop improved decision making processes has been explicitly linked with its alleged value in admitting a wider range of popular wants into the political calculus*. This implies that for their own ultimate benefit people should be ready to specify their wants more and more precisely for the benefit of the policy analyst, the psychologist, or risk assessor. Yet if we stand aside from this activist or positivist framework, a different perspective suggests itself. What if people's wants appear inconsistent and confused to the external analyst precisely because one of their central wants is to keep their options open and informal, to be free to revise and renegotiate their priorities and objectives in a social context that they find meaningful and open to their own influence? Kekes is one of the few political philosophers to have entertained something approaching this possibility (Wynne 1975, 1981):

> "The (dominant) assumption is a bias in favor of an activist theory of rationality. By this I mean that it is supposed that people have some goals which they desire in order to gain what they regard as benefits, and rationality is what maximizes their chances (overall) of getting

---

express a sense of alienation and complete "otherness" to the process whereby technology is developed and implemented. As I have argued elsewhere, it is understandable that "the masses" should see technology in these terms, as penetrating from outside (from a world of the technological elite) which is almost as alien as Martian decision making. See, for example, Wynne (1980), Winner (1978), and McDermott (1977).

*See, for example, Pearce's proposals for elaboration of public inquiry procedures in Britain (Pearce 1979). Like motherhood, it is hard to criticize the wish to accommodate more political wants in decision making, but the approach objectifies the notion of wants and thereby ignores power relationships.

> the goods. This activist way of thinking about rationality is not so much mistaken as incomplete...
>
> There are two sources of change in human affairs. One is the attraction of ends to achieve, and activist theories of rationality would do justice to this; the other is the necessity of avoiding problems, dangers, and difficulties. Avoidance behaviour is no less rational than a goal-directed one. But such behaviour may consist in doing nothing, or acting so as to make sure that something will not happen, or its success may reside in pre-empting dangerous options, or making sure that conditions for future performances are not removed. This passivist approach is not directed towards the achievement of any particular end. Its function is to guarantee that the conditions exist in which ends can be pursued".

The activist response could of course be that this only expands the list of possible goals that could be actively sought. However, this response ignores the fact that the more we move into the realm of such "defensive" goals, the more fluid and ambiguous they become, and the more they evade any rationalistic requirement of precision. Far from being a sign of irrationality, however, this very informality and inexplicitness of goals can be seen as a sign of a very robust form of rationality, even if it is incompatible with that of professional analysts and political elites. It may be the realm of ambiguity and vagueness of goals that is their defense against inflexible and risky control by external professional groups and alien elites. In this respect it is interesting that concern about the health effects of the nuclear industry frequently focuses upon the use in principle of cost-benefit approaches in setting discharge levels, and exposure limits. Pro-nuclear advocates are not slow to point out the irrationality of this attitude when implicit cost-benefit trade-offs are made by those same critics regularly in daily life. However, seen as an attempt to defend that very freedom to make *flexible* trade-offs, amenable to local, familiar, and informal social influence, this hostility towards formalized, externally controlled cost-benefit balancing appears rational enough.

Just as the rationalist model of science assumes that the principles and reasons for scientific judgments can be formally, fully, and precisely specified, so the same perspective expresses a faith in our ability and need to specify our goals and values precisely and formally. The alternative perspective on scientific rationality sees it as a more informal, open-ended and flexible social achievement, leaving scientists free to revise and renegotiate judgments as they go along, in their own familiar special cultural setting.

Once this view of scientific rationality is accepted, it is easier to accept that people should rationally attempt to retain their freedom to hold ambiguous goals, and to leave some goals and values unexpressed because as such they are richer, more amenable to flexible social negotiation and development in thier own ordinary terms, and thus less vulnerable to external appropriation and control. External groups who attempt (with however pure democratic intent) to assist such groups by insisting that their wants be specified more precisely for the purposes of policy making only polarize the social bifurcation even further by consolidating the sense amongst their own culture that people "out there" are obstinately perverse or obscure in their attitudes whilst at the same time consolidating the sense of social threat amongst the "analyzed" of external, alien, and inflexible control by the formalized expropriation of more and more of their shrinking realm of autonomous symbols of social intercourse. As I have said elsewhere, a "dual society" structure is consolidated on the basis of this general rationalistic mythology which appears to be every bit as strong and as damaging as the social cleavages between "developers" and "peasants" in Third World contexts, and where by far the most significant risks are those social ones to do with the extent to which people should yield up control of their existence to unknown alien groups—be they "development experts" usually called decision making elites, or development experts in the conventional sense—whose values and objectives they do not understand and cannot therefore be expected to trust (Wynne 1975, 1981). If, as I believe, this is the root problem surrounding risk assessment—a problem of legitimacy in the most basic sense—we will not begin to tackle it without examining the social structure of technological innovation and the myths that defend it.

7. CONCLUSIONS—SOME ISSUES FOR RISK ASSESSMENT RESEARCH AND PRACTICE

In this paper, I have attempted to outline some of the implications of a model of science which has dominated and pervaded political and professional discourse, especially where scientific methods have been deliberately employed in political decision making. I have tried to delineate an alternative account of scientific rationality that has been developed from a wide range of detailed case studies and analyses of science as actually practised over the last ten or more years. At least in Europe this account has gained the status of the orthodoxy amongst sociologists, historians, and philosophers of science. Whether it ever gains wide acknowledgment remains in question just because it challenges myths and ideologies built around the rationalist image of science, which are central foundations of power structures and processes in advanced industrial society. These images have the effect of creating the conditions of political consensus by default of active intervention. Of particular relevance in the present context is that they weave an impenetrable social barrier round the process where real commitments are made to different social futures through technology, and present technology from this encircled, elite concalve as if objectively given in natural or historical law. In any case, policy analysts who take seriously their own claims to be open always to new truths however difficult, should recognize the force and the implications of the different account of scientific rationality that is gaining currency. If this causes us to treat risk assessment as an integral part of political anthropology generally, that would be a sign of progress. In the light of this general perspective certain research issues can be identified as important to pursue.

(i) There is a continuing need to research the social-psychological undercurrents of risk perceptions in various technological fields, but with two major provisos in mind. The first is that technology has to be conceptualized primarily as a social organization entity, not a physical entity—it therefore has *intrinsic* political dimensions not merely external political possibilities (Winner 1980). Thus the insight that "risk" itself may often be a category of thought inserted artificially by risk assessors into people's minds can be further developed to address directly the political question of how people "on the receiving end" regard decision making systems that develop and control technology—what factors shape their sense of alienation or trust, why, and so on. In doing this we move away from the focus on physical risks as a baseline, and instead take the more realistic line of seeing attitudes to physical risk as a variable conditioned by social and political contexts and perceptions. Determining objective physical risks will still be valid of course, but the lingering tendency to start from this scientific vantage point and add social perceptions as qualifications to the "objective" physical picture must be completely reversed. A second proviso to this point is that the ethos of research so far in this area has been psychological. This has provided insights of great importance. But there is a need to complement this perspective with social anthropology, to locate the kinds of attitude revealed in actual life situations, to understand their validity to the people concerned in *their* social circumstances, not in those of a cossetted elite. This being said, one can see how energy risk perception studies might be extended into new technologies such as information and biotechnologies since the factors affecting attitudes have been identified and could generate several interesting hypotheses for technological fields different from energy. I have outlined some of these in a separate research proposal (Wynne 1981b).

(ii) Seeing technology as a social entity, and scientific rationality as intrinsically conditioned by social commitments, allows us to identify a new area of possible risks. Whereas previously dominant concepts have helped maintain the assumption that accidents have been the consequence of *sporadic* lapses of concentration, discipline, skill, etc., a more soicological perspective requires that we examine the social-psychological and structural-organizational features that systematically generate risky attitudes. Even theterms available to describe the syndrome—dogma, arrogance, complacency—evoke an individualized, moral sense of responsibility (hence sporadic failings) rather than systematically generated processes that occur even in the most exquisitely rational of institutions*. Thus the sociology of science has demonstrated the inevitable and *positive* role of institutionalized dogma and a blind faith in eventual success in scientific research. It is necessary as a means of channelling collective intellectual effort and sustaining morale and concentration on a demanding long-term project. But it is of course ambivalent in that it can also lead to the neglect or arrogant dismissal of significant anomalies or alternatives. Whereas in pure science this may not matter in the sense that physical risks do not hang directly on the issue, in a technological organization great physical risks may arise. (In the case of "technological" sciences such as recombinant DNA research the same type of risk will also prevail.) There is thus a great deal of scope for research on organizations responsible for the development and control of technology, so as to illuminate the formal and informal social processes that may systematically generate complacency, dogma, and related risky attitudes and practices. At the same time those very processes of myth elaboration which generate risky attitudes also generate social polarization and the dual society syndrome. Conversely, there is a need to examine the effects upon morale and concentration (and thus upon riskiness and viability in an opposite sense) of initiatives to establish "insider" pluralistic criticism as a safeguard against such tendencies. There is a social-psychological "catch-22" here that has yet to be explored either at a theoretical or a case study level. One would require participant observation and theoretical sophisitication on a par with the Latour and Woolgar (1980) internal, detailed anthropological study of laboratory life.

(iii) The social interpretation of uncertainty in the technology politics field is an area overdue for rigorous and sustained treatment. The issue arises implicitly or explicitly, in every case study or discussion of risk assessment. Yet no coherent, cumulative analysis has been developed. A general question around which such a program could be constructed would be related to the role of tacit images of science in such processes as I have discussed earlier—to what extent and in what specific ways is conflict surrounding the implications of technology interpreted by those involved as the result of residual (or chronic) ignorance? An influential example is Weinberg's idea that some issues such as low-level-radiation risks are trans-scientific in that they can be defined in the abstract as scientific questions, yet in practice can never be resolved**. On the sociological

---

*Here the perspective on scientific rationality which I am advancing parts company with that of other theorists such as Simon and March. The latter have recognized "bounded rationality" in organizations, but the implication is that those boundaries can be continually enlarged in a single system of rationality. Implicit in this model is an ideal type, perfect rationality derived from the positivist image of science, which acts as a heuristic framework (March and Simon 1979). See also LaPorte (1980) and Mayer (1971).

**See Weinberg (1972). Kemeny (1979) expressed the same assumption when he observed "There was a time when if someone asked me for information on science I'd say, 'Go and ask a scientist'. And if they said, 'Which one?' I probably would have replied 'As long as you get a good one it doesn't matter which one, because if the question lies within the scientist's area of expertise and he or she investigates the problem

account of science I have advanced, one can say that this is only a polar example of what is always the case, in that there is never *inescapable* empirical proof of any scientific knowledge, but social agreement amongst scientists that things are so. The implication of Weinberg's trans-science analysis is that if the scientific issue could be finally resolved then there would be no more conflict. But this is almost tautological in that science is defined in such a way as to require that conclusion. Relating to the previous point for policy research, where I emphasized that scientists were for good reason regularly incapable of defining uncertainties and anomalies in their own knowledge, it would be valuable to research the ways in which scientists involved in risk analyses and policy debates interpreted the nature and degree of certainty (and relevance) of their knowledge in different contexts of debate, and to see how others responded. A topical example here is the recent case of the expert Lowther Committee in the UK, which investigated the effects of environmental lead. Quite apart from many other interesting aspects of the affair, one expert member of the committee is widely regarded as having expressed incompatible views on a relevant scientific matter, in the committee's account of it and his own account in a scientific paper published at the same time. This points us towards the hitherto unrecognized extent to which scientific knowledge is *always* incomplete in meaning, and open to very different interpretations in different contexts of use. Pinch has demonstrated this process in pure science, and the same process applies in other areas*. This also hints at the related point that science is *always* less formally defensible against scepticism than is usually acknowledged (witness the endless elaboration of the fluoridation risks issue) and what is legitimate and illegitimate interpretation and insistence on insignificant uncertainty is a socially conditioned matter. Whether the relevant social conditions are within science or go beyond science is an open question in every case, and needs case-by-case empirical investigation. The point of the sociological perspective on rationality in this area can be illustrated by reference to the widely cited paper on risk perception by Slovic and Fischoff (1980) "How safe is safe enough?"**. In this paper the authors discuss several ways in which entirely objective perceptions and expressions of risk are not achieved. Exaggerated certainty, resistance to contradictory evidence, and other items mentioned before, are also identified by Slovic and Fischoff. The crucial difference, however, is that they do not seem to recognize any distinction between individual psychology and institutional processes, and the emphasis upon the former has the effect of supressing any acknowledgment of the latter. This again implies that the failings of objective perception are virtually confined to the unempirical "peasantry", the errors of the decision making/analytical elites being isolated and occasional, not structurally inevitable in the social structure of technology and science. Slovic and Fischoff also place their faith in the call for the experts to explicate their premises and heuristics more clearly. Yet this is also part and parcel of the positivist myth of scientific rationality which I am criticizing. In this case their perspective is not so much dismantled as transcended into a new set of questions.

---

long enough, the scientific truth will be obtained'. I still believe that, but it's not always of practical value because some of the technology we use is at the very frontiers of scientific knowledge." In other words, conflict only arises because one scientist or another is unqualified, or because the research hasn't been developed enough.

*A suggestive case study, although in a science out of the political pressure of the policy area, is Pinch (1981).

**See also Slovic *et al.* (1974). Tversky and Kahneman (1974) performed their work with scientists, but the sociological implications of their findings were not discussed.

(iv) A general theme that pervades the previous ideas for research and indeed the whole paper is the assertion that a particular false image of science has become deeply embedded in thought and debate in the policy field, as a rhetoric of authority. The most important general research task is to seek out those unwarranted assumptions and *a priori* formulations embedded in the concrete language of such issues. This will naturally reflect upon the role and tacit languages of risk analysts themselves, and will inevitably cause us to examine the social origins of persistent perceptions of "risk" (i.e., conflict) in power relations in society. Especially relevant to our starting point of technology will be the question of who enjoys power in the real processes of technological commitment, as opposed to that of who takes part in formalized decision making processes that often appear to be of marginal relevance to the real power structures. To analysts with anthropological leanings like myself it is a matter of course that political order will always rest upon myths which are effective to the degree that they are not directly challenged. Myths about an order corresponding to scientific rationality have played their part in sustaining social authority, but I have suggested that the rationalistic self-delusions of decision making elites and their policy analysts may now be inadvertently polarizing and deepening the schisms and political instabilities of industrial society.

In other words, before we attempt to decide what specific research and practical objectives should be adopted in the risk analysis field, we should perhaps decide whether our own role is to perpetuate these myths or to identify and explicate them. Is our analysis to be social research or social therapy? If the latter, are we sure that the therapy will not be iatrogenic?

REFERENCES

Barnes, S.B. (1972) Sociological explanation and natural science: A Kuhnian reappraisal. European Journal of Sociology 13:273-93.
Barnes, S.B., and Shapin, S.A. (Eds.) (1979) Natural Order. London: Sage.
Boguslaw, R. (1972) The New Utopians. Chicago: Chicago University Press.
Burn, D. (1979) Nuclear Power and the Energy Crisis. London: MacMillan.
Cameron, I., and Edge, D.O. (1977) Images of Science. London: Butterworth.
Collins, H.M. (Ed.)(1981) Knowledge and controversy: Studies of modern natural science. Social Studies of Science 11:February.
Conrad, J. (Ed.)(1980) Society, Technology and Risk Assessment. London: Academic Press.
Doty, P. (1972) Can investigations improve scientific advice? The case of the ABM. Minerva 10:280-94.
Douglas, M. (1970) Natural Symbols. Hardmonsworth: Penguin.
Douglas, M. (1975) Implicit Meanings. London: Routledge.
Edelman, M. (1966) The Symbolic Uses of Politics. Illinois: University of Illinois Press.
Edelman, M. (1971) Politics as Symbolic Action. New York: Academic Press.
Edelman, M. (1977) Political Language: Words that Succeed and Policies that Fail. New York: Academic Press.
Ezrahi, Y. (1974) The authority of science in politics, in Science and Values edited by A. Thackray and E. Mendelsohn. New York: Humanities Press.
Firth, R. (1973) Symbols: Public and Private. London: Allen and Unwin.
Forman, P. (1971) Weimar culture: Causality and Quantum Physics. Historical Studies in the Physical Sciences 3:1-115.
Gellner, E. (1974) The Legitimation of Belief. London: Cambridge University Press.
Habermas, J. (1975) Legitimation Crisis. Boston: Beacon Press.
Hesse, M.B. (1974) The Structure of Scientific Inference. London: MacMillan.
Hoos, I. (1969) Systems Analysis and Public Policy. California: University of California Press.

Kemeny, J.G. (Chairman)(1979) The President's Commission on the Accident at Three Mile Island. Washington, D.C.
Kemeny, J.G. (1980) Saving the American democracy: The lessons of Three Mile Island. Technology Review, June/July:65-75.
LaPorte, T. (1980) Design and management of nearly error free safety, in Social Science Aspects of the Accident at Three Mile Island, edited by D. Sills. Colorado: Westview Press.
Lathrop, J. (1980) The Role of Risk Assessment in Facility Siting: A Case Study from California. WP-80-150. Laxenburg, Austria: International Institute for Applied Systems Analysis.
Latour, B., and Woolgar, S. (1980) Laboratory Life: A Study in the Social Construction of Scientific Facts. London: Sage.
McDermott, J. (1977) Technology: Opiate of the intellectuals, in Technology and Man's Future, edited by A. Teich. New York: St. Martin's Press.
McKenzie, D. (1981) Statistics and Society, 1850-1920. Edinburgh: University of Edinburgh Press.
Majone, G., and Quade, E.S. (1980)(Eds.) Pitfalls of Analysis. Chichester: Wiley.
March, J.G., and Simon, H.A. (1979) Organisations.
Mayer, M.W. (Ed.)(1971) Structure, Systems and Symbols: Readings in Organizational Behavior. Boston: Little Brown.
Mazur, A. (1977) Science courts. Minerva 15:1-13.
Merton, R.K. (1975) Sociology of Science. Chicago: University of Chicago Press.
Mitroff, I. (1974) The Subjective Side of Science. Elsevier.
Mulkay, M.J. (1980) Science and the Sociology of Knowledge. London: Allen and Unwin.
Nelkin, D. (1971) Nuclear Power and Its Critics. Ithaca, N.Y.: Cornell University Press.
Nelkin, D. (1981) Some social and political dimensions for nuclear power: Examples from TMI. American Political Science Review 75:April.
Niblett, C. (1979) Episodes in the History of Industrial Science: Three Case Studies. Ph.D. Thesis, University of Manchester.
Okrent, D. (1980) Comment on societal risk. Science 207:372-5.
Otway, H., and von Winterfeldt, D. (1980) Beyond Acceptable Risk: On the Social Acceptability of Technology. London: Academic Press.
Pearce, D. (1979) Social cost-benefit analysis and nuclear future. Energy Economics 1:66-71.
Pinch, T.J. (1981) The sun-set: The presentation of certainty in scientific life. Social Studies of Science 11:131-58.
Ravetz, J.R. (1974) Scientific Knowledge and its Social Problems. Harmondsworth: Penguin.
Robbins, D., and Johnson, R. (1976) The role of cognitive and occupational differentiation in scientific controversies. Social Studies of Science 6:165-87.
Scheffler, I. (1967) Science and Subjectivity. New York: Bobbs Merrill.
Slovic, P., and Fischoff, B. (1980) How safe is safe enough? Pages 121-47 in Risk and Chance, edited by J. Dowie. Milton Keynes: Open University Press.
Slovic, P., Fischoff, B., and Lichtenstein, S. (1976) Cognitive processes and societal risk-taking. Pages 165-84 in Cognition and Social Behaviour, edited by J.S. Carroll and J.W. Payne. New York: Wiley and Halstead Press.
Sperber, D. (1975) Rethinking Symbolism. London: Cambridge University Press.
Starr, C., and Whipple, C. (1980) Risks of risk decisions. Science 208:1114-29.
Tversky, A., and Kahnemann, D. (1974) Judgement under uncertainty: Heuristics and biases. Science 185:1124-31.
Weinberg, A. (1972) Science and trans-science. Minerva 10;209-22.
Weizenbaum, J. (1980) Computer Power and Human Reason. Boston: MIT Press.
Welsh, I., and Wynne, B. (1981) Nuclear power and political control—The Windscale fire as a case study in symbolic uses of science. Paper presented at the British Society for the History of Science Conference on New Perspectives in the History of Technology, Machester, April.
Williams, R. (1980) The Nuclear Power Decisions. London: Croom Helm.
Winner, L. (1978) Autonomous Technology. Boston: MIT Press.

Winner, L. (1980) Do artifacts have politics? Daedalus 109:121-35.

Wynne, B. (1975) The rhetoric of consensus politics: A critique of technology assessment. Research Policy 4:1-51.

Wynne, B. (1980) Technology, risk and participation, in Society, Technology and Risk Assessment, edited by J. Conrad. London: Academic Press.

Wynne, B. (1981a) Nuclear Decision Making—Rationality or Ritual? Ch.7. London: British Society for the History of Science.

Wynne, B. (1981b) Decision Making, Risk, and Information Technology—An Outline for Research. University of Lancaster.

Wynne, B. (1981c) Redefining the question of risk and public acceptability—the social viability of technology. Paper presented at the EEC FAST Programme, February.

Zebraski, E. (1980) Lessons learned and lessons not yet learned in nuclear safety. Paper presented at AAAS Meeting, San Francisco, January, p8.

A Proposal to Create a Cultural Theory of Risk

Michael Thompson and Aaron Wildavsky
International Institute for Applied Systems Analysis, Laxenburg, Austria

1. INTRODUCTION

Perhaps the greatest achievement of anthropology has been to show that different people, faced with the same situation, do different things. No sooner does an economist, a psychologist, a sociologist, or a political scientist produce his universally valid model of some aspect of human behavior than an anthropologist will jump up and say, "Ah, but what about the Bongo-bongo?" It is probably safe to say that there is no universalisitic proposition that is immune to Bongo-bongoism*. Some semiologists, for instance, have suggested that there are some signs so rooted in nature that their meaning *has* to be intrinsic. One such sign is the arrow: →. Here, surely, the meaning is intrinsic to the sign; arrows always fly through the air point first and so the point of the sign, surely, must always indicate the direction of travel. Quite so, but who said the sign was an arrow? On a remote island in Micronesia the people spend much of their time hunting a secretive bird whose feet are specially adapted to the marshy terrain. Each time it puts one of its three-toed feet down on the ground, it leaves a sign: → : and every time a hunter sees one of these signs, he knows with certainty which way to go to catch up with his quarry. He goes in the direction indicated by the big central toe.

Anthropologists have been so carried away by their spoilsport success that they have almost lost sight of the one really interesting question which is: given that different people in the same sort of situation do different things why do they do the different things that they do? This is the question that political culture tries to answer. If it was only the people on remote Micronesian islands who did things differently then political culture, whilst intellectually, intriguing, would be of little practical relevance, but though we might like to pretend otherwise, this is not the case. The simple but unpalatable fact is that the Bongo-bongo are alive and well and living right here in our midst.

More than a century ago General Booth, the founder of the Salvation Army, after describing the iniquities perpetrated by the slave traders in the unknown African interior, suddenly brought the whole outrage uncomfortably close to home and asked: "Is there not also a darkest England?" In much the same way, the cultural approach brings the Bongo-bongo home to roost; the only difference is that this time it is our rationality, not our morality, that is outraged**.

The trick with cultural contexts is to come at policy-relevant debates, such as those that surround technological risk, from a contrary direction. Instead of asking "What are the risks?" we ask "What would you like the risks to be?" Though the analysis of cultural contexts has something useful to contribute to all public policy analysis, its point of entry is that state of desperation, exasperation and exhaustion that is reached when, after years of debate and after the expenditure of millions of dollars, we are still no nearer agreement on what the risks out there are.

---

*Credit for this fictitious tribe, for the far-from-fictitious "ism" that it highlights, and for much of the development of the cultural contextual approach is due to Mary Douglas. Particularly relevant are Douglas (1972, 1979).

**Though, as we will argue presently, morality and rationality are closely related.

1.1. An Unconventional View of Culture

Although it is commonly assumed that to reject cultural universalism is to embrace cultural relativity, this assumption would only be valid if the number of different things that people could do was infinite. Political culture* starts out from the intuitive hunch that, though people faced with the same situation do different things, there aren't *that* many different ways of doing things. This idea that there are *patterns of culture*, that such patterns are accessible and describable, and that there are not very many of them is not a synthesis of the universalistic and relativistic traditions; it is a rejection of them both. It is as well to make this point explicit right from the beginning. Political culture is born of the marriage of anthropology to political science, and culture is its key concept, but the idea of culture that underpins this approach is far removed from the ideas that have held conventional sway in anthropology.

We start to lose interest in those vague pseudo-entities like, say, American culture or French culture; instead we start to focus on the various *cultural biases* that are to be found (in varying proportions) within both American society and French society. Nor do we persist in our sympathy for either of those contradictory formulations that would have us believe, on the one hand, that culture is just some kind of reflection or rationalization of social action, and on the other, that it is some kind of rule book for the game of social life that gets handed down, largely unchanged, from generation to generation. Culture, we begin to feel, conforms to neither of these contradictory extremes of total fluidity and rigid concreteness. Culture is plastic. Although it can be pushed this way and that way, it cannot be pushed just anywhere; and just to push it into some fresh attainable configuration (and then keep it there) requires a great deal of social effort. Instead of a lot of social actors programmed day-in-day-out by culture and instead of some cultural superstructure that forms like a fluffy cloud above the granite mountain of production and consumption forces, we have mentally and physically creative individuals for whom culture is a rather provisional thing that needs to be made afresh (or, at the very least, patched-up and modified) each morning.

Our hypothesis is that there are only a few shareable (i.e., socially viable) cultural biases and that (in advanced industrial societies, at any rate) it would be most unwise to assume that any of these possible biases is "uninhabited". Each bias is stabilized, albeit precariously, by its distinctive *cosmology*. By cosmology is meant those shared beliefs and convictions about how the world is that sustain and justify moral judgments. But, if shared beliefs and convictions are rather provisional things that have to be worked at all the time, how can you be sure that yours are in line with everyone else's? Only by engaging in the continual process in which each individual justifies his own actions and passes judgment on those of others can you become, and remain, a member of a moral community. "Giving a good account of yourself" is not just a figure of speech; it is the human propensity that makes human society possible.

The final piece of the cultural jigsaw is *social context*, for it is an individual's social context, so the theory runs, that determines his particular cultural bias and leads him to give credence to one cosmology—to one set of shareable beliefs about how the world is—rather than to any of the contradictory alternatives. And finally, social context, it is held, is adequately described by just two dimensions: *group*, which has to do with the extent to which the individual is involved in bounded social groups, and *grid*, which has to do with the extent to which the individual is involved in hierarchical arrangements (either of individuals or of bounded groups of individuals).

If you are convinced that the world is like this, and I am convinced that the world is like that, then we are likely to act in very different ways in the one physical world that we both inhabit. That is, people who subscribe to contradictory

*Our cultural approach adds to, rather than contradicts, that of the economist. Where political economy looks at how people are differentiated by constraints—by what they *can't* do—political culture looks at how they are differentiated by capabilities—by different things that they do with what they *can* do.

cosmologies will operate contradictory rationalities. Each, ensnared in his own provincial rationality, will see the others as the Bongo-bongo (Lord Rothschild 1974, for instance, sees the Friends of the Earth as "eco-nuts" and "eco-maniacs" and they for their part see him as something quite unprintable). Since each is committed to the rightness of his own cosmology, neither can let go of it long enough to concede the validity of the other's rationalities. In the acrimonious battle between these rationalities no one can afford to stand outside it; to do that would be to concede the validity of other rationalities and to admit that any approach aimed at determining which rationality is right is bound to be wrong, and the whole misguided purpose of the battle is not to live with the Bongo-bongo but to annihilate them.

Cultural contexts constitute the systematic deprovincialization of rationality. If individuals in different social contexts are firmly attached to contradictory convictions about how the world is, then it is only to be expected that they will have very different ideas of what the risks are out there. To ask "Who is right?" is not just to ask a question that probably only history can answer; it is to encourage the arbitrary tyranny of one provincial rationality over all the others. The intellectual source of this arbitrariness we call "the individualist fallacy".

1.2. The Individualist Fallacy

The decision maker, when he finally arrives at the realization that he is faced, not with a technical problem but with one that has to do with people, tends to reach for his psychologist. Wrong again! The individualist fallacy is a specific version of the fallacy of misplaced concreteness (Whitehead 1926). In the first instance, this fallacy takes the form of assuming that risk is something inherent in our external world—that the risks that threaten us are determined by the inherent physical properties of the universe. It is this assumption that leads the decision maker to believe that the problem he faces is a technical one* and that leads him to hire various technical experts who claim to be able to solve it. Two difficulties result. First, the technical experts can *never* gain access to all the risks that are determined by the universe. Second, public policy involves the public, and some of its members it turns out do not revise their varied perceptions of the risks "out there" to bring them into line with those of the experts.

Which social choice should the decision maker take: the Platonic one based on what the real risks are (as far as the expert can see, that is) or the Benthamite one based on some aggregation of what people believe the risks to be? If he chooses the Benthamite alternative then he will need to hire some different experts—those who can tell him what people think the risks are. But the psychologist still retains as his datum the real risks "out there". He discovers threshold points for low-probability/high-consequence events, he discovers persistent overestimators and persistent underestimators, optimists and pessimists,...,risk-accepting and risk-averting individuals; all plotted in against the same vast expanse of misplaced concrete—that totality of risks determined by the universe.

Far from being dismayed by the unattainability of what the risk assessor and the psychologist have set their hearts on, the anthropologist will point out that this unattainable goal is not what risk is about anyway. Of course, the universe is not irrelevant, but it is not what determines the area of concern that we are referring to when we talk about risk. Rather than being something that is inherent in the external world, risk and its absence are qualities that are *conferred upon* it by social processes. These social processes, as they blot out some risks that are really there and as they set down others that have no counterpart in physical reality, create a fluctuating pool of risks somewhere between us and the universe. Since any debate about risk must take place within a social setting, it will inevitably be a debate about the properties of this fluid pool. Anyone who claims that

---

*Rather than one that has some technical *aspects*. This is not to say that the technical expert has no part to play in risk management, only that, when he insists that the problem itself is technical he is playing the wrong part.

it is not—that it is about the concrete expanse that lies somewhere beyond this pool—is falsely claiming to be a "cosmic exile" (Quine 1960, p275)*.

This—the cosmic exile's impossible claim—is the fallacy of misplaced concrete ness. It is the naive assumption of the literalist who, when he looks at the world, believes that he is seeing it with the naked eye. It is a serious fallacy. It is not some trivial little objection to be circumvented by conceding a touch of "subjectivity" as the engineer moves from analysis to evaluation; nor is it something that can be put right by the addition of a "social science input"—by bringing in "the behavioral engineers" to sort out the "people problems". It is nothing less than a total misapprehension of what risk is.

Risk, though it has some roots in nature, is inevitably subject to social processes. Since we (being members of society) are at one end of these processes, we can never gain access to the raw unprocessed reality. Whether we like it or not, the risks to which we have access are *processed* risks. If only we concede that this is so, and stop pretending that we can get at the risks before they have been processed, then we can begin to understand something about them—we can begin to understand the processes of which they are the end products. Also the first thing that we must understand is that these processes have very little to do with the individual as an isolated entity; they demand very little of his innate sensory apparatus but a great deal of his socially acquired referential apparatus. They are preeminently *social* processes.

An individual is led to impose cut-off points, not because his eyesight is not good enough or his nose not up to the job of sniffing beyond a certain range, but because of the social and cultural institutions that are stabilized and made credible for him by virtue of his social context and its appropriate cosmology. As long as those institutions remain credible—as long as he is prepared to go along with them—they will do the risk management (the imposition of thresholds and the setting of their levels) for him automatically. A different individual in a different social context is led to impose on his external world all kinds of risks that have no physical counterpart (or at least none that is detectable), not because he is suffering from some serious malfunction of his sensory equipment, but because the different social and cultural institutions that he finds credible do it for him, automatically.

Human life has transferred risk perception from the individual (as a psychophysiological unit) to the social fabric of which he is part (and as an adaptive mechanism it has, up to now, proved remarkably successful). In their handling of risk humans do not act as individuals but as social beings sensitively tuned to social pressures and submissive to mutual coercion.

1.3. The Individual as a Social Being

The hypothesis in terms of the individual and his social context is specifically designed to handle the individual, not as an isolated entity, but as a social being. The social units that do the risk handling come in a variety of forms—bounded groups, hierarchical organizations, competing personal networks..., atomized communities**—and they run the entire gamut from vast federal agencies to tiny self-help arrangements organized by nothing more formal than a shared sense of neighborliness. The two dimensions of social context allow us to go behind these contingent differences and to categorize any individual according to the way in which he is involved with, or free from, these various social units.

If he belongs to a bounded group that can impose severe sanctions on its members then he will score positively on the group dimension. If he is at the center, of an extensive personal network he will be imposing prescriptions upon those individuals who are towards its periphery and so he will score negatively on the grid

*A situation sympathetically described by an economist who himself exhibits the bias appropriate to this context. See Friedman (1980).

**This is the one that by and large, has been missed by social scientists. See Munch (1974) and Thompson (1981).

dimension*. If he wishes to purchase some drug that has been banned by the Food and Drug Administration his life will be limited (ever so slightly) by an impersonally imposed prescription. If he is subject to many such prescriptions (if he is peripheral to the personal networks of others, for instance) he will score positively on the grid dimension. On the other hand, if he is central to an extensive network and immune from group loyalties and sanctions, he will have the resources, the necessary information and the will to just go out and buy the forbidden drug on the grey market**. In that case he will score negatively on both group and grid dimensions. If he is a self-employed and largely self-sufficient farmer whose involvement with his similarly situated neighbors, though convivial, is essentially voluntary then he will score zero on both dimensions.

In this way, as we plot our individuals onto the social map, we build up a scatter diagram that will reveal which sorts of social units predominate in that society, and highlight where the potentially troublesome polarizations of affiliation to those units are located. Individuals in one category of social context will, as social beings, be sensitively tuned (by their cosmology and their strategy) to the social pressures characteristic of that context and they will be disposed to submit themselves to the kinds of mutual coercion that are characteristic of that context.

These different kinds of social pressure and different kinds of acceptable coercion manifest themselves in social and cultural institutions. The Sherpa, by and large, avoids mentioning the names of the dead; the lower-caste Hindu, by and large, defers to the high-caste Brahmin; the lineage member whose crops have been spoiled by flooding goes to his kin on the higher ground and asks them to make good his loss and they for their part, mindful that in periods of drought positions are likely to be reversed, accede to his request. In other words, different kinds of institutions are appropriate to different social contexts. Social context is, as it were, the soil in which institutions grow. The institutions (changing the metaphor) are a kind of automatic pilot; the individual grants credence to them and, in return, the institutions look after the risks for him. From this it follows that

(i) different kinds of institution will tend to flourish in different social "soils"—that different institutionalized ways of handling risk will evolve in different social contexts—and

(ii) that whether an institution flourishes or withers will depend on whether individuals continue to grant credibility to it.

If an individual's social context, for some reason, changes then he will be tempted to override the automatic pilot that previously handled his risks for him. He will begin to question the legitimacy of the institutions, and risks of which previously he was scarcely aware will suddenly, as they become his personal concern, loom large and threatening.

If we looked only at the institutions, and not at the individuals who either support or fail to support them, then we could say nothing about the appropriateness of institutions nor could we begin to understand the dynamic social processes that distribute credibility this way rather than that and, in so doing, uphold one institution and cause the collapse of another. For instance, perhaps the most alarming of all the problems that have emerged from the nuclear debate has to do with the way in which institutions that have long given sterling service have become paralyzed. Our institutions are designed to provide us with decisions (and with good decisions, to boot) yet everywhere we see policies stymied and nuclear industries declining into bankruptcy, not because our institutions have decided that all things considered that is what should be happening, but because they simply can no

*There are good theoretical reasons why the contradictory situations in which he is both included and excluded by groups and in which he is central to one powerful network and peripheral to others are unlikely to occur. Nor do they seem to occur in practice. See Thompson (1981).

**A stituation sympathetically described by an economist who himself exhibits the bias appropriate to this context. See Friedman (1980).

longer come up with *any* decisions—even bad ones! Any approach that concerns itself only with the institutions and not with the social dynamics of their support will be powerless to explain why hitherto healthy institutions have suddenly become paralyzed in this distressing way. And, if it cannot diagnose the disease, what hope is there that it will be able to come up with prescriptions that will effect a recovery?

2. SOCIAL CONTEXT

In the social sciences, psychologists have traditionally focused on the *individual* while sociologists and anthropologists have concentrated on the larger-scale individuals—clans, classes, lineages, hierarchies, age-grades,...,corporations. Both these—the individual and the various relationships that make him into a member of society—can be handled by the concept of social context. It is a concept that, crossing these customary disciplinary boundaries, allows us to move smoothly back and forth between the micro-concerns of the psychologist and the macro-concerns of the sociologist and anthropologist. The totality is described in terms of each of its constituent elements (individuals) *and* their various social contexts (their myriad relationships with other individuals)*.

But the usefulness of this social context concept will depend on the extent to which it can make some generalizations about the differences between individuals. If at the end of our scrutiny each individual emerges with a unique social context that distinguishes and sets him apart from every other individual, then little theoretical or practical progress will have been made. The same is true for those other extremes of outcome: that in which every individual (being related to every other individual) emerges with an identical social context, and that in which we cannot distinguish any criteria for deciding whether one individual's social context is the same as or different from that of some other individual. On the positive side, if these obstacles can be avoided then social context will provide us with a concept applicable to any human being, anywhere... anytime—it is not subject to cultural**, historical, technological, ecological or social qualifications. Whether it is of any use or not will depend entirely on whether it can avoid being so particularistic that every individual has to be seen as a special case without, at the same time, being so universalistic that everyone ends up the same.

Our hypothesis, in fact, maps distinct cultural categories into our two dimensions of social context (see Figure 1). This mapping in which the two dimensions of social context permit us to distinguish five distinct categories into which individuals must fall, each distinct conjunction of individual and social context being stabilized by a distinctive *cultural bias* (or *cosmology*), most certainly avoids these twin pitfalls: particularism and universalism. Five not only lies somewhere between one and infinity, it is an eminently handleable number as well. If we only need take five biases—five kinds of social individual, five cosmologies,

---

*The individual, it should be stressed, is handled as a social being, not as a unique psycho-physiological entity. But it is not being argued that individuals are indistinguishable empty vessels until they are filled with the breath of social life, only that the manner of their involvement in social life will superimpose a distinctive bias upon whatever was there to begin with. Once an individual has become a social being it is often difficult, and sometimes impossible, to say where this line between original content (nature) and social overlay (nurture) lies. A consequence of this is that, to the extent that this line is blurred, the concerns of the psychologist and the anthropologist overlap. Anthropology has often tended to pull back from this fuzzy region; the present approach does the opposite.

**Though we have spoken of it as a cultural approach it is, properly speaking, an approach in terms of cultural *bias*. We are interested in *patterns of culture* rather than in culture itself and so, in this sense, the concept of social context is free from cultural qualification.

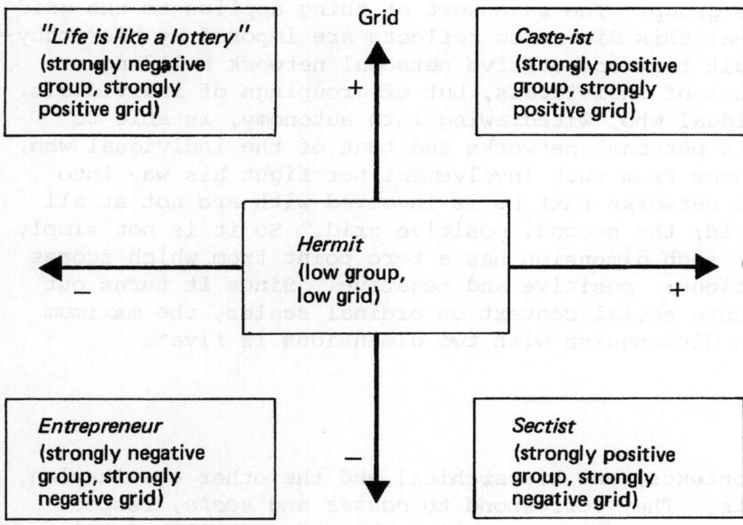

Figure 1. Social context and cultural categories.

five kinds of perceptions of risk,..., five risk-handling strategies—into account then an anthropological theory of risk is not just possible, it is usable as well*.

2.1. The Two Dimensions of Social Context

The *group* dimension can be visualized as running from "individualized" to "collectivized", but the way in which it varies independently of the *grid* dimension (which can be visualized as running from "egalitarian" to "hierarchical") is not so obvious. Grid has to do with the extent to which an individual is subject to (or free from) socially imposed prescriptions. For example, the member of a high caste is clearly a member of a bounded social group, but the dietary and pollution-avoiding prescriptions that he has to observe are not the consequences of his membership of the group itself but of that group's relationships with other groups—of its rigorously defined and energetically maintained position within a hierarchical arrangement of groups. By contrast, a member of an egalitarian group—one with no internal differentiation and with no defined relationships with other groups—would score high on the group dimension but low on the grid dimension.

But this idea of high and low scores on the two dimensions is still not enough to capture the full complexity of what is going on. The social context of an individual who is not a member of any group, because there are no groups around for him to be a member of, is very different from that of the individual who finds himself expelled from the group to which he looks in vain for his support. The first has

---

*We should stress here that the commitment is not to the number five but to *intermediate sociology*; to the existence of some handleable and cross-culturally valid basis for the disaggregation of individuals into a number of categories, that number lying somewhere between one (universalism) and infinity (particularism). For reasons of practicality and application, we would prefer the number to be small rather than large. But we would have no objections to this intermediate number being, say, three or six or eight if such a categorization gave better results in practice and could be predicted theoretically by making certain specified changes in the initial assumptions (and provided those changes were more realistic than those used to generate the fivefold categorization. Where Merton (1968) advanced sociology by directing it to middle-level propositions, we would like to shift sociology and anthropology onto a modest number of cultural contexts.

zero group; the second, negative group. The same sort of thing applies to the grid dimension. The prescriptions that this dimension reflects are imposed by hierarchy—either the hierarchies that result from competitive personal network building or the hierarchical arrangements, not of individuals, but of groupings of individuals. Again, the context of the individual who, withdrawing into autonomy, is able to avoid any coercive involvement in personal networks and that of the individual who, try as he may, can neither withdraw from such involvement nor fight his way into any central positions within the networks that he is involved with are not at all the same. The first has zero grid; the second, positive grid. So it is not simply a matter of high and low scores; each dimension has a zero point from which scores may be high or low in two directions: positive and negative. Since it turns out that it is only possible to measure social context on ordinal scales, the maximum number of categories that we can distinguish with two dimensions is five*.

2.2. The Five Contexts

The two strongly grouped contexts, one hierarchical and the other egalitarian, are well known to anthropologists. They correspond to *castes* and *sects*, respectively. We free these terms from the settings to which they are usually restricted (the anthropology of the Indian subcontinent and the sociology of religion, respectively) and use them to describe two distinct cultural biases that we observe in Western secular society—*caste-ist* and *sectist***.

A *sect* erects a wall of virtue between itself and the nasty outside world from which it wishes to set itself apart. The members collectively *reject* the outside world—they do not negotiate any sort of relationship with it. The result is that, though the collectivity may exercise almost total control over its members, it can do nothing to the rest of the society.

A *caste* separates itself off, not with a wall of virtue, but by means of clearly defined distinctions between it and those other groupings that exist outside it. The result, as each caste defines itself by its distinction from (yet clearly specified interrelation with) other castes, is a complex hierarchical framework of status distinction, prescriptions, restrictive practices, correct channels, and proper procedures. The members of a caste, therefore, do not reject the outside world; they collectively take up a clearly specified position *within* it. A caste, as a result, can come to exercise a high level of control over its own members *and* over those outside its boundary.

The top left context is where we find the social condition that, intuitively, we associate with poverty. Very alone and everywhere hedged about by externally imposed prescriptions, the individual in this context has little control over events in time and space and his (or her, since women tend to predominate in this context) cosmology tends to be cobbled together from such ill assorted bits and pieces as he can lay his hands on. This is the social context of *kitsch*, of millenarianism, of inconsistent eclecticism and, above all, of Lady Luck as the provider and withholder of all resources. Engels, in his account of the condition of the un-unionized working classes in Victorian Britain, is describing this context and so too (though with a slightly different set of prescriptions in mind) is Mrs Gaskell when she speaks of those for whom "life is like a lottery" (see Engels 1845, Gaskell 1958, 1971).

The bottom left context is where we find the entrepreneur—the individual who is often blamed for the sorry predicament of the "life is like a lottery" man. Unlike the sectist and the caste-ist, the entrepreneur has no interest in the maintenance of transactional boundaries; he profits from their demolition. Free from group obligations and sitting at the center of an extensive personal network of his own creation, he is not just free from prescriptions—he is, as a result of his

---

*Strictly speaking, the maximum number is nine, but it turns out that when the third dimension (manipulation) is introduced, only five out of these nine have stabilizable equilibria associated with them (see Figure 2).

**This distinction has been operationalized in terms of 13 criteria for caste-ist and sectist groups. See Thompson (1980a).

economies of scale and his willingness to treat anything (even his grandmother) as negotiable, actively imposing prescriptions on those less forceful individuals who form the periphery of his network. He is a pragmatic materialist.

The fifth context is the one that is often missed by occidental social scientists; it is the distinctive context occupied by the individual who has deliberately chosen to keep his involvement in socially binding relationships (be they network or group relationships) to a minimum. This is the social context of the hermit—the still center of the social hurricane (but not the raucous North American pseudo hermits like Thoreau, who are really sect leaders in search of followers). Nor are such individuals always *outside* society. Though they may choose to withdraw from all social involvement, it is only *coercive* involvement that they have to avoid to stabilize themselves in this context. Whole convivial societies can be stabilized around this context with the help of its "live and let live","sufficient unto the day is the evil thereof" cosmology (see Thompson 1980b, 1981a,b).

2.3. Power and Control

Each conjunction of social context and cosmology will generate its own distinctive strategy. That strategy will result either in the individual manipulating others or being himself manipulated. The members of a sect end up manipulated (collectively); the members of a caste (collectively) manipulate. Entrepreneurs clearly are (individualistic) manipulators while those whose lives are "like a lottery" are equally clearly being manipulated (individually). Only those individuals who operate the autonomous strategy appropriate to the hermit, and who successfully avoid all coercive involvement with their fellow men, will end up neither manipulating nor manipulated*.

If we add this third dimension, *manipulation*, to the two social context dimensions and join these five equilibrium states together in the simplest possible way we obtain a graph like Figure 2. Topologically, if you have two basins and two hilltops arranged in this way then the landscape must contain a fifth equilibrium

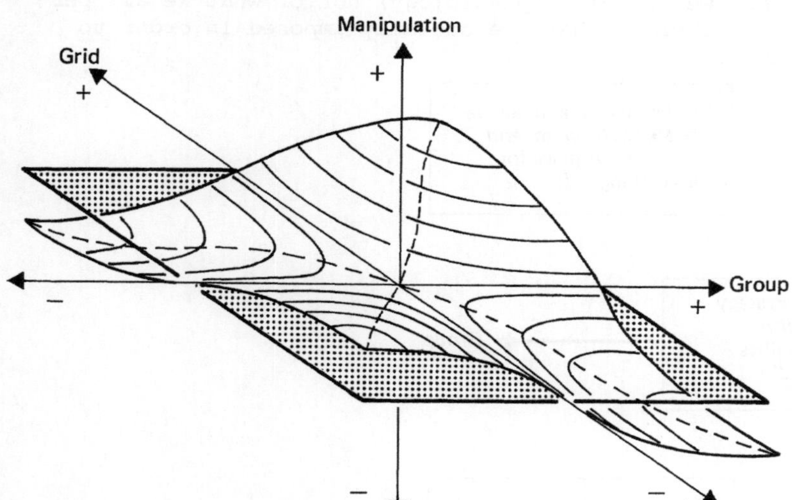

Figure 2. Relationship between social context and manipulation.

*For some, the word "manipulate" may be value-laden. No such judgment is intended here. The simple fact is that social life often involves manipulation—a fact that some will deny and others will acknowledge. The present argument is that these divergent responses (and the further distinctions between those who, in acknowledging this fact, regard manipulation as inevitable, regrettable, avoidable..., desirable) are not uninfluenced by social context.

state—a saddle point. It may, of course, contain other equilibrium points as well but this is its simplest possible configuration. The five equilibrium states (the five "flat bits") are now clearly separated from each other by means of disequilibrium states (the intervening slopes). This diagram (or simple variations of it) constitutes the form of our hypothesis. Its content follows.

2.4. The Five Strategies

With each social context there goes a distinct rationality—a world view, a cosmology, a cultural bias..., a particular way of seeing the world and man's place within it—that provides a moral justification for certain kinds of actions and a basis for moral disapprobation of other kinds of actions. The idea is that social context and world view will tend to stabilize one another and that, as people in a shared context come to share a particular world view, they acquire and sustain a particular morality that enables them continually to make judgments on human actions: rewarding some and punishing others. The hypothesis states that such shareability—such stabilization of moral community—can only occur at or near these five equilibrium states. Each cultural context requires a distinct personal strategy for the individuals in it to maintain its stability. Individuals in different social context will tend to home in onto distinctive strategies that will enable them to steer an optimal personal course through all these socially generated rewards and penalties. If you observe an individual as he follows one of these strategies, you will discover whether he ends up manipulating others or being himself manipulated (see Figure 3).

2.5. The Social Bases of Perception

The combination of world view and strategy that is appropriate to each social context results in an individual in that context perceiving his external world in a distinctive way. Thus our hypothesis forms the basis for an anthropological theory of perception*—not of how we perceive (physiology) nor of what we all perceive (psychology) but of the *patternings* that are socially imposed in order to support certain moral commitments.

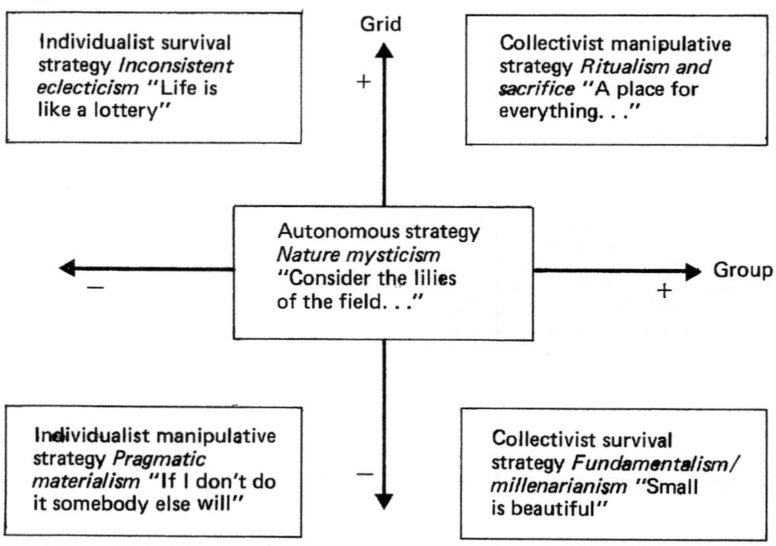

Figure 3. Thumbnail sketch of the strategies, the cosmologies, and the justifications.

*"Perception" is being used here in the lay sense of gut convictions about how things are rather than in the narrower meanings of the word as it is defined by psychologists and physiologists. See Otway (1979).

Risk is very much a moral question (and never more so than when it is being asserted that it is not). As the history of risk assessment clearly demonstrates, widely divergent convictions as to what the risks "out there" are can often coexist within the same society. Our hypothesis, we argue, is capable of handling these cultural biases both as to how risks are perceived and as to how they are evaluated. The current state of this theory, which we wish to develop further, is discussed next.

3. THE CULTURAL THEORY OF RISK

Are risks perceived in the short or long run? Making just the simple division of time into the short term and the long term is sufficient to separate three contexts—those in which we will find experts—from the other two—those in which experts are absent. Expertise and perception of the long term go hand in hand; so, if you only perceive the short term, there is no chance of your becoming an expert. But your inability to perceive the long term may be by choice or it may be by compulsion, and this distinction provides a second criterion that enables us to separate these two expert-less contexts—that of the hermit (by choice) and that of the "life is like a lottery" man (by compulsion). Turning to the three categories in which the long term *is* perceived, we meet three different kinds of expert and the problem is to explain why each kind of expert is appropriate to his particular category.

Both the entrepreneur and the sect member are able to perceive both the short and the long terms but they evaluate them very differently. For the entrepreneur, the short term dominates the long term; he is in the business of manipulation but he is realistic enough to know that his manipulation, being the product of his own energy (rather than of the authority of an institutionalized office that he, for a time, fills), does not extend too far into the future. Being an expansive optimist, he allays his fears that his short-term successes may not continue indefinitely by insisting that the long term will turn out to be a prolongation of the short term. He is predisposed to give credence to the "business as usual" scenario.

The sect member's evaluations of the short and long terms are the reverse of the entrepreneur's. Collectivized within his wall of virtue, and with little control over the short term, his main concern is just to survive; he sees himself as one of the meek who, in the long term, will inherit the earth*. In this way, the optimistically perceived long term comes to dominate the gloomy short term. If there is to be a long term at all then the short term will have to be radically changed now. He is, in consequence, predisposed to grant credence to the "no growth" (the "radical change now") scenario.

Where both the entrepreneur's and the sect member's evaluations of the long and short terms are unbalanced (with the short term dominant for the entrepreneur and the long term dominant for the sect member) the evaluations of the caste member are quite nicely balanced. This is the context occupied by the planner and the bureaucrat. Insulated from the pressing daily concerns of the entrepreneur by the institutionalized framework that guarantees the continued existence of the office that he fills, he is able to give adequate attention to the long term. What is more, he sees events in that long term as being controllable, not by him personally, but by the complex collectivity of which he is a self-effacing part. Being part of an elaborate hierarchy he is predisposed to be sensitive to fine distinctions and, in consequence, is unlikely to see the long term as a mere extension of the short term or *vice versa* (and, of course, he would be out of a job if he conceded that there was no such distinction). Each is seen in a balanced and discriminating way and, since collective control over events is seen as extending far beyond the short term, the long term is viewed with cautious optimism. The result is a willingness to grant credence to the "middle of the road" scenario**.

*Not surprisingly, this friend of the earth often joins with others to become a Friend of the Earth and together they draw up "Blueprints for Survival" designed to ensure that the earth does not disappear before their inheritance falls due.
**For instance, see Schanz (1979). For further evidence of the separation of these

We can summarize these criteria—long- *versus* short-sightedness, choice *versus* compulsion, short-term *versus* long-term dominance, and balanced *versus* unbalanced evaluation—for separating the five categories with the help of the basic diagram, Figure 4.

3.1. Acceptable Risk

How does the acceptability of risks vary with social context? There are some philosophical problems here because, as we have just argued, the perception of risks also varies with social context. Just because an individual cannot see a risk it does not follow that he is not exposed to it (and an individual in a different social context may well be able to see that he is exposed to it). But, if he does not even know that the risk that he is exposed to is there, is it valid to speak of him "accepting" that risk? On top of this there are the risks that, though they are not actually there, are believed to be there (and an individual in another social context may well be able to see that they are not there). Such nonexistent risks may, in some contexts, constitute a major proportion of the risks that are perceived in those contexts and they may be managed by all sorts of socially imposed rewards and penalties which, in turn, provide the incentives for their acceptance or avoidance.

The hypothesis is designed to cope with these sorts of problems but, before we talk about "the acceptability of risk", we should bear in mind the fact that the pool of risks to which our acceptance/avoidance criteria will be applied is itself highly fluid. It may fill up or empty according to whether our perception extends to the long term or is restricted to the short term, and it may fill up with all sorts of risks that aren't there but are believed to be there. For instance, there is no real physical risk involved in eating cooked rice on which the shadow of an untouchable has fallen (or, at least, no more risk than if the shadow had not fallen on it), yet such a risk is believed to be there, and a great deal of trouble is taken to avoid it, in the strongly positive group and strongly positive grid context of the high-caste Hindu. The risk, of course, is to the intricate and highly discriminating social fabric but it is externalized and given expression in terms of the physical world of light and shadow, nutrition and bodily processes.

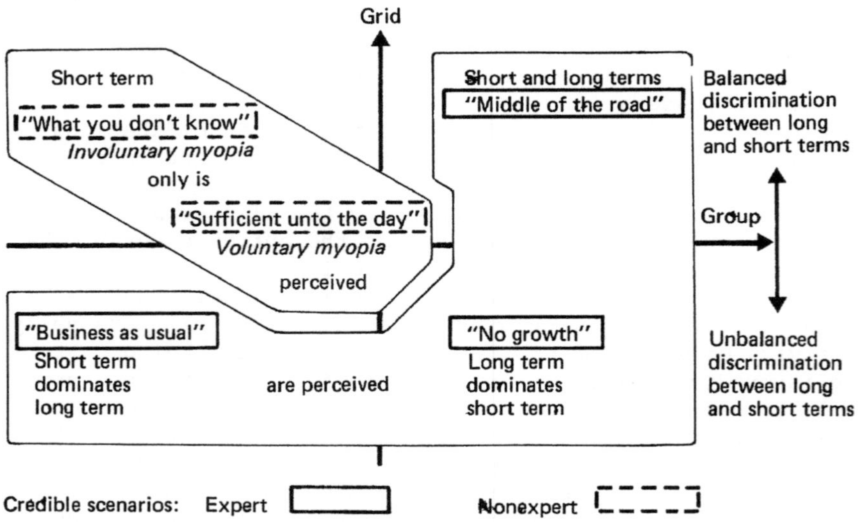

Figure 4. Criteria for separating the five categories.

---

three scenarios, in the form of a matrix showing how the other incredible scenarios are perceived from each credible scenario, see Reuyl *et al.* (1977).

Pollution concepts develop rather differently in the social context of the sect member because his group is internally undifferentiated and rejects, rather than negotiates relationships with, the rest of society. Here there is only one crucial boundary to be defended—that which separates and protects the good vulnerable "us" on the inside from the nasty predatory "them" on the outside. Only by ceaselessly patrolling and maintaining this boundary can the sect retain its cohesion and survive through time. So the risks are real enough and the institutionalized steps that are taken to minimize them—the witchcraft accusations, the denunciations, the confessions, and the explusions—do serve a vital purpose but, though they are expressed as such, they are *not* risks inherent in the physical world*. And, even when the risk is there in the physical world, the concern that surrounds it may well derive from some physically nonexistent *social* risk that overlies it. Let there be no doubt that many of the most feared risks in modern technology are of this social kind. The cultural contextual approach provides us with a means of identifying them *and* of taking them seriously.

The way in which all these factors—(1) the perception of risks, (2) their acceptability, (3) the overlying of social (physically nonexistent) risks, and (4) the rewards and penalties for different kinds of risks—vary with social context can be summarized as in Figure 5.

3.2. Styles of Risk Management

What sorts of cultural contexts lead to different styles of risk management? In the "life is like a lottery" context, individuals are alone and powerless to influence events. Such institutions as are able to emerge here furnish the occupants, not with a style for managing risks, but rather for just absorbing them. They make something of a virtue out of necessity. Since the occupants have ignorance imposed upon them, they might as well see it as blissful and, since they have no way of mitigating whatever the risks out there are, they might as well comfort themselves by the tough minded and slightly braggardly assertion that "what you don't know can't harm you". To have to search for and identify the cause of every harm that befell the members of this context would divert their meager resources away from their number one concern: survival.

Yet this seeking of an explanation for every ill is precisely what those in the other survival context—that of the sect member—insist on. In this context a person dies, not because "his number is up", but because someone somewhere has caused him to die. It may be his own transgression or it may be the work of some other agent and the whole institutionalized framework of social risks and their causality is invoked in order to find the culprit and to exact the appropriate penalty. The easy give-and-take that concepts of chance and probability bring to the interplay between harm and its causation is not to be found here. No "background" risk is acceptable, all harm has to be accounted for... partial models full of slippage, tolerance and expediency and incorporating the idea that some risk is inevitable, and that particular deaths are its statistically inevitable outcome, are rejected. Instead, we have a holistic style of risk management—a total system model in which causal links can be (and are) traced until the blame for every particular misfortune can be laid at some particular door. The anthropological literature on witchcraft accusations (the classic work being Evans-Pritchard 1937) clearly reveals that these linkages are always traced in such a way as to minimize, not the physical risks, but the social risks.

Why should these "home-made" theories of causation and blame be so very different between these two survival contexts? The answer is that elaborate theories of causation and blame can only be constructed within a scaffolding of social risks and social risks can only exist if there is some social structure there for them to threaten. For the sect, its wall of virtue is its *raison d'être*; in the atomized setting of indivualist survival there *is* no social structure.

*The reason why they have to be expressed in the medium of real physical risks is that, under the strict regime of collectivized control that tends to be generated by this context, this is just about the only permissible medium. See Owen (1981).

(1) Only short-term risks perceived with any clarity.
(2) No options as to acceptance or rejection. Risk acceptance a "fact of life".
(3) No social boundaries to maintain so little concern for pollution and therefore few, if any, nonexistent risks.
(4) Personal rewards seen as resulting from luck not personal risk taking and they are not resented.

(1) Short- and long-term risks perceived. Long term seen as different from short term but controllable.
(2) Risk averting. If risks cannot be avoided completely they are spread.
(3) Many boundaries and distinctions to be maintained so elaborate pollution concepts and high incidence of nonexistent risks.
(4) Personal risk for personal gain penalized. Personal risk for totality's gain rewarded.

(1) Only short-term risks perceived with any clarity.
(2) If their avoidance would require social involvement (e.g., risk spreading through reciprocity and social obligations) then acceptance is lesser of the two evils.
(3) Little concern for pollution so few, if any, nonexistent risks.
(4) Personal risk taking for personal reward approved of as long as it does not involve coercion of others.

(1) Long and short terms perceived. Dominance of short term over long term maintains expansive optimism.
(2) Risk as opportunity.
(3) Little concern for pollution (entrepreneurs profit from removal of social boundaries).
(4) Personal risk for personal reward approved (even if it causes coercion). Personal risk for benefit of totality less popular.

(1) Short and long terms perceived. Concern for survival causes long term to dominate short term.
(2) Strong aversion to all risks except those involved in the defense of wall of virtue.
(3) Pollution concerns all clustered around a single social boundary and give rise to many nonexistent risks.
(4) Zero-sum mentality penalizes the personal risk for personal gain. Only risks taken for the totality are rewarded (often posthumously).

Figure 5. Risk and social context.

The styles of risk management that emerge in the context of the hermit and the entrepreneur have a number of features in common. Since both contexts are individualized, and since both focus optimistically on the short term, there is a tendency to regard risk management as a personal business and to emphasize the importance of individual skill and judgment. Where there exists a choice between handling risks individualistically (by the market, for instance) and handling them collectively (by regulation, for instance) both hermit and entrepreneur will favor the former. Both will be biased against institutions that collectivize risks, or convert voluntary into involuntary ones. They will instead tend to give their support to those more diffuse institutions that, directly or indirectly (but more likely indirectly), increase the areas of risk that are left to individual values and decrease the areas that are handled by social choice.

However, despite these similarities, there is some divergence in risk handling style which derives from the different levels of manipulation that go with these two contexts. The hermit neither coerces others nor is himself coerced and this means that, when he supports institutions that individualize risks, those risks remain closely attached to the individual who chooses to take them. But, when the entrepreneur supports similar institutions, his risks are, to some extent, exported to those individuals who he is successfully manipulating. This is undoubtedly a very effective, if inequitable, way to manage risk and much of our present worries about risk are attributable to the fact that many risks that used to be exported in this way are now bearing down upon individuals in other, less impotent, contexts. As has often been pointed out, the difference between the risks involved in coalmining and in nuclear power is that in the former they kill only coalminers while in the latter they may kill you and me as well.

In the two contexts with strongly positive group—castes and sects—there will be a bias toward institutions that take risk handling away from the individual and give it to the whole, or to some agency charged with the responsibility for handling it on behalf of the whole. The internally undifferentiated (and therefore egalitarian) sect will favor holistic risk management; the internally differentiated (and therefore hierarchical) caste will favor management by specialized and profesionally staffed agencies.

Such hierarchical systems are based on many fine yet clearly defined distinctions and they build up into complex social structures that are all too easily threatened. This means that social risks—anything that threatens to diminish the clarity of these distinctions—are particularly to be feared, and the result is that risks come to be handled not holistically but in a compartmentalized way. It is this chopping up of risks (and everything else) in order to minimize the social risks that justifies the cruel definition of bureaucracies as systems incapable of learning from their mistakes. Yet their record for handling social risks, albeit of their own making, is vastly superior to that of the sects. Sects are always falling apart but bureaucracies seem to know intuitively how to ensure their own existence—they do it so instinctively that they scarcely make any mistakes to learn from. Nor, when it comes to the real physical risks, is any purpose served by urging bureaucracies to abandon their chopped-up models in favor of holistic ones—it is not in their nature to do this. What is possible is to reorganize some of the dividing lines, even making some new ones where appropriate, so that as the social risks are managed the physical ones get looked after as well. And, of course, it is always possible (it has just been done for energy in the United States) to create a new high-status agency specially charged with the task of cutting through the boundaries created by other lower-status agencies within the framework.

It is significant that the study of risk (which has largely been developed in and for this sort of context) has itself been compartmentalized. Risk assessment has been restricted to the real (physical world) risks while social scientists have been careful to restrict themselves to the social (the physically nonexistent) risks. Both are reluctant to step outside their spheres of competence and put the two kinds of risk together.

3.3. The Efficacy of the Five Styles

Having tentatively described these different styles of risk management, how effective are they?

A first temptation might be to assume that, since the whole debate is conducted in the idiom of real physical risks, those styles that concern themselves only with those risks will be more effective than those that concern themselves with risks that are, in fact, nonexistent physically. But who is to say which are the more potentially damaging: the physical risks "out there" or the real physical consequences of the collapse of a large part of our social order and its stabilizing institutions? Rather than rush to such hasty and simple minded evaluations of these different styles of risk management, we should concede that in social systems where such threatenable social structures exist (and that includes *all* advanced industrialized nations) social risks are among the most serious of the risks that have, somehow or other, to be managed. It is no good crying that physical scientists cannot be expected to give their attention to social risks, nor can the social scientists justify their remaining inside their disciplinary stockade. Experts may divide the world up into areas of expertise, responsibility may be chopped up between government departments, and select committees may be limited by their terms of reference but the fact remains that the risks we face and the risks we have to manage are not neatly compartmentalized; like it or not, they spill across from technology to technology* and from the physical world into the social world.

In adopting a style of risk management based on the assessment of the external (real physical world) risks, we come more and more to resemble the members of W.H. Auden's expedition:

> ...sound on Expectation
> Had there been situations to be in;
> Unluckily they were their situation...
>
> (From *The Quest*.)

How can we modify our risk management so that it becomes reponsive both to the external and the internal risks?

A first step is the recognition that both kinds of risk are there, and the second step is the development of some sort of theory that casts its net wide enough to catch them both. This is what the cultural approach is aimed at. It generates sets of hypotheses that predict how and when such risks will be present and which allow us to recognize them for what they are. From the debate we can disentangle the external and the internal risks—the physical and the social—and we can take each kind seriously. When we recognize that we have caught a social—a physically nonexistent—risk in our methodological net, we do not say, "Oh, it's just a social risk" and throw it back into the ocean. We recognize that it is a risk—a particular kind of risk—and we can refer it to our conceptual scheme to obtain some estimation of how serious a risk it is and of how it might best be handled.

4. RELEVANCE FOR PUBLIC POLICY

With the development of a cultural contextual theory of risk, policy makers will know three things they do not now know:

(i) why there are profound disagreements over risk;
(ii) why disagreements between certain groups in specific contexts cannot be reconciled; and
(iii) with what kinds of groups in which contexts it is possible to come to an accommodation.

Risk is always a social product and the cultural theory of risk will help the policy maker to handle it as such.

---

*For instance, from nuclear weapons into nuclear power. There are two kinds of ratioactive material in the United States—military and civilian. This is not a physical distinction—it is an administrative one—and it is a distinction that is probably not shared by anyone wishing to steal such material in order to make a nuclear device of his own. The Nuclear Regulatory Commission is currently funding a research project (at Lawrence Livermore Laboratory) to consider systematically all the possibilities of such diversion but, because its terms of reference do not extend to the military material, the crucial calculation of the point at which as the security system is improved the adversary will decide that it is easier to steal a ready-made bomb rather than all that civilian material that has been completely overlooked.

REFERENCES

Douglas, M. (1972) Environments at risk, in Ecology: The Shaping Enquiry, edited by J. Benthall. London: Longman.
Douglas, M. (1979) Cultural bias. Occasional Papers of the Royal Anthropological Institute, London.
Engels, F. (1845) The Condition of the Working Class in England (translated 1958). Oxford: Blackwell.
Evans-Pritchard, E.E. (1937) Witchcraft, Oracles and Magic Among the Azande. Oxford: Oxford University Press.
Friedman, M. (1980) The government as nanny. The Listener, April.
Gaskell, E. (1848) Mary Barton.
Gaskell, E. (1855) North and South.
Merton, R. (1968) Social Theory and Social Structure. New York: Free Press.
Munch, P.A. (1974) Anarchy and anomie in an atomistic society. Man 2:243-61.
Otway, H.J. (1979) The perception of technological risks. Paper presented at the CEC/CERD Seminar on Technological Risk: Its Perception and Handling in the European Community, Berlin, April 1979.
Owen, D. (1981) Spectral evidence: body metaphor and social experience in the Salem witchcraft trials, in Essays in the Sociology of Perception, edited by M. Douglas. London: Routledge and Kegan Paul.
Quine, W.V.O. (1960) Word and Object. Cambridge Mass.: MIT Press.
Reuyl, J.S., Harmon, W.W., Carlson, R.C., Levine, M.D., and Witwer, J.G. (1977) Solar Energy in America's Future. Stanford, California: Stanford Research Institute.
Rothschild, Lord (1978) Risk. The Listener, November.
Schanz, J.J.Jr. (1979) Oil and gas reserves: welcome to uncertainty. Resources 58: March.
Thompson, M. (1980a) Fission and fusion in nuclear society. Rain (newsletter of the Royal Anthropological Institute), December.
Thompson, M. (1980b) The aesthetics of risk: culture or context, in Societal Risk Assessment, edited by R. Schwing and W. Albers. New York: Plenum.
Thompson, M. (1981a) The problem of the center, in Essays in the Sociology of Perception, edited by M. Douglas. London: Routledge and Kegan Paul.
Thompson, M. (1981b) A three-dimensional model of risk perception, in Essays in the Sociology of Perception, edited by M. Douglas. London: Routledge and Kegan Paul.
Whitehead, A.N. (1926) Science and the Modern World. New York: MacMillan.

Chapter 4

DECISION PROCESSES AND PRESCRIPTIVE ASPECTS OF RISK

Evaluating Technological Risk: Prescriptive and Descriptive Perspectives

John W. Lathrop

International Institute for Applied Systems Analysis, Laxenburg, Austria

1. INTRODUCTION

Decisions concerning the deployment and management of novel or hazardous technologies raise several issues involving the evaluation of their impacts on society. Examples of such decisions include the siting of a liquefied natural gas facility, the regulation of nuclear energy production, and the screening and regulation of toxic chemicals. Each of these kinds of decisions results in uncertain benefits and costs to society. It would seem reasonable, then, that such decisions could be aided by any of several analytic techniques, including cost-benefit analysis, or perhaps decision analysis, which could include in the evaluation attitudes toward uncertainty and value trade-offs between conflicting objectives. However, there are often special aspects involved in such decisions that can make standard technical or economic analyses not very useful for aiding political decision making processes. These aspects include outcomes of the decision having very serious negative consequences with very low probability, inequitable distribution of burden, large scale, novelty, and others to be discussed below. Decisions involving such aspects sometimes come to be known as problems in managing social risk. Even though the word risk is currently in wide use in the media, it is often defined or applied in different ways by different parties for the decision at hand. In spite of this serious problem, to be discussed at some length below, the need to appraise the risks presented by a new or hazardous technology has led to the development of several analytic techniques often referred to collectively as risk assessment. Yet those techniques generally assume, either implicitly or explicitly certain prescriptive objective functions that are not sensitive to important societal concerns about potentially hazardous technologies. As a result, such techniques may be useful as inputs to a political decision making process, but are not as helpful as they could be if they used objective functions more descriptive of the relevant social concerns. This point is illustrated by cases where a member of the technical risk assessment community assumes a simple objective function, such as minimizing expected lives lost or life expectancy lost, observes individual or government behavior that does not minimize that objective function, then suggests that therefore something is wrong with the decision makers involved (see, e.g., Rothschild 1979). That deduction is not the only one that could be made from the evidence. It could equally well be deduced that the objective function is inadequate.

The technical risk assessment community is not the only set of people addressing problems of social risk management. There is a growing body of research that is developing descriptions and explanations of human behavior that does not minimize narrow technical objective functions (Kunreuther 1980). Yet the descriptions and explanations most sensitive to individual attitudes are not oriented toward developing broader social objective functions that can be directly used to aid the political decision making process in managing social risk. The development of evaluation

---

*The research reported in this paper is supported by the Bundesministerium für Forschung und Technologie, FRG, Contract No. 321/7591/RGB 8001. While support for this work is gratefully acknowledged, the views expressed are the author's own and not necessarily shared by the sponsor.

models and techniques by the technical and psychological research communities forms a rough spectrum from the narrow but readily applied technical objective function to the broad set of social and psychological considerations that are not readily applied to aiding the decision process. This paper seeks to identify the part of that spectrum that, if further developed, may be the most useful in that it may provide usable decision aids that are also sensitive to the important societal concerns involved.

2. THE BASIS OF THE PROBLEM: TWO PERSPECTIVES OF RISK

While the previous section referred to the range of approaches to risk evaluation as a spectrum, it has underlying it two poles of thought which, in turn, underly the theme of this paper. The basis of the problem of managing technological risk is that it is not a single concept, but is viewed from two distinct perspectives: the technical and the psychological.

2.1. Technical Perspective

From the technical perspective, technological risk is some probability distribution over sets of negative effects. Those effects are often limited to health effects, or casualties. For example, Keeney (1980b) defines the risk of an action as the probability distribution over sets of individual probabilities of dying (p_i, $i = 1,\ldots,N$), where p_i is the probability that the ith individual will die before the end of the next time period due to the action taken. The probability distribution over those sets is required to represent dependencies between the fatalities. Other analysts use summary measures of that distribution. In the Rasmussen report (Reactor Safety Study) the risk of a nuclear reactor is represented by a reverse cumulative probability distribution over numbers of fatalities per year per reference reactor (USNRC 1975). That distribution is sometimes referred to as a Rasmussen curve. Other risk assessments have used the same measure (Hazelwood and Philipson 1977). Other analysts go one step further and reduce the distribution to the expected number of health effects (e.g., Pate 1978). While it would be perhaps unfair to say that such analysts define risk as expected number of health effects, the fact remains that the negative consequences of a technological option are measured by that index.

There is one definition of risk not discussed here, and that is "risk is probability times consequence". This is because this discussion is limited to representations of the risk of decision alternatives. Probability times consequence may describe the risk of an event, but could not in general describe the risk of any decision alternative with more than two possible outcomes.

2.2. Psychological Perspective

In contrast to the fairly straightforward risk definitions listed above, extensive psychological research in the field of risk perception has suggested many more dimensions to be included in a definition of risk (Slovic *et al.* 1980, Linnerooth 1978, Kunreuther 1980). While these dimensions are covered in some detail in a later section, they are listed and briefly discussed here as a way of describing technological risk from a psychological perspective. They are listed roughly in order from the most easily adapted to a simple technical measure to the most difficult considerations to measure. While this discussion limits itself to health effects for brevity, there is no particular reason not to include other effects: financial, quality of environment, etc.

2.2.1. *Expected Number of Health Effects*

While this aspect was listed above as part of the technical perspective, that does not mean that it cannot be a part of a psychological perspective also.

2.2.2. *Possibility of Catastrophe*

Effects can be spread out over time and space, such as car fatalities, or they can be "bunched" into a catastrophe at one place and time, such as a possible major nuclear accident. That bunching can be very important for how society views the risk. Effects can also be bunched by cause, by identifiable population group, or by state of nature, as will be explained later.

2.2.3. *Equity*

There are actually two related aspects of equity that are important in the perception of risk. The first is the amount of overlap between the populations at risk and at benefit. Clearly if a technology benefits none of those at risk, that is a case of inequity. But there is another, more subtle aspect of inequity that has to do with the ease of identifying the population at risk. If a technology can be identified as negatively affecting 100 people in the US, while benefiting all citizens, that may be seen as a necessary evil. But what if those 100 are all poor, or all asbestos workers? That is a matter of greater concern.

2.2.4. *Degree of Control*

This aspect is a more general version of the voluntary/involuntary distinction made by Starr (1969). The central concept here is the level of participation of each potential impactee in each of two decisions: to expose himself to the risk, and to deploy the technology in the first place.

2.2.5. *Attributability*

This is an important aspect of social reaction to a risk that is often overlooked because of the cause-specific way risk analyses are done. It could be that generally incompetent engineering in cars kills far more people than a particular gas tank design. Yet the more easily identified cause of accidents, the gas tank, may give risk to a much stronger public reaction.

2.2.6. *Nonprobabilistic Factors*

Perhaps the most serious mismatch between technical and psychological perspectives lies in the evaluation of risk related to severe outcomes with no regard for the estimated probability of those outcomes. While there may be good reasons for that sort of evaluation, it can cause serious problems in developing consistent decision aids.

2.2.7. *Nondecision Comparisons*

The search for criteria for acceptable risk often falls back on comparisons not involved in actual decisions. For example, comparisons are often made between some technological risk and moving to Denver, or smoking an extra cigarette, or driving an extra mile. Yet very few people choose between living near a nuclear reactor or living in Denver. Perhaps even fewer choose between living near a liquefied natural gas terminal and smoking an extra cigarette, though that decision involves some complicating factors.

2.2.8. *Nontechnological Factors*

Some aspects of a technology having little to do with possible health effects may have a great deal to do with perceived risk. Those aspects include the degree of centralization, the extent to which deploying a technology would infringe on civil liberties, how closely a new technology is linked with a high-consumption lifestyle, etc. While these aspects seem far removed from evaluating the health effect risk of a technology, they should be included in any effort to understand possible opposition to a technology. Such aspects may be as important or more so than any of the other aspects listed above in determining differences in what levels of risk are held to be acceptable.

2.3. Results of the Two Perspectives

The differences between the two perspectives presented above are very relevant to many problems in technological risk management. Technological risks are typically evaluated using analyses that assume the technical perspective on risk. Yet very often the political process in which the decisions actually get made is responsive to the psychological perspective. As an example of this problem consider the case of the selection of a site for a liquefied natural gas import terminal in California. A technical risk analysis by a competent technical consulting firm found a proposed site at Oxnard, California, to be very safe, with very low risk to the community. Yet that analysis stated that a maximum credible accident (MCA) could involve up to 70 000 fatalities, though only by a seemingly incredible series of events that could only occur with an extremely low probability. In the political process, however, consideration of such large numbers of fatalities led to a requirement that the terminal be sited remotely, away from any city, precluding the Oxnard site (Ahern 1980). The applicant made plans according to a technical risk analysis, then had to change plans as a result of the political process.

As the example just presented illustrates, one result of the gap between the technical and psychological perspectives of risk is that technical risk analyses are often not effective in the political process. Conversely, another result is that the political process often does not receive the help it could from the scientific and engineering community.

Another result of the two perspectives on risk is that the community of risk analysis scientists and engineers work with analysis tools spanning a broad spectrum. One end of that spectrum involves prescriptive evaluation models that offer very clear guidance to decision makers, but that are not sensitive to psychological concerns. The other end of the spectrum involves psychologists and other social scientists identifying societal concerns very well, but not in a manner that results in clear guidance to the decision making process. Clearly, what is called for is more development of risk evaluation models that are sensitive to social and psychological concerns, yet are meaningful and useful aids to the political decision making process. Section 4 below spells out several considerations for that model development. Before that, however, the next section briefly reviews two existing approaches to risk evaluation that bracket the part of the analysis spectrum proposed for development in Section 4.

3. TWO APPROACHES TO RISK EVALUATION

3.1. Multi-attribute Utility Functions

Within the framework of decision analysis, multi-attribute utility theory can be used to develop essentially prescriptive risk evaluation models that take into account more than the expected number of health effects. As one example, Keeney

(1980b) sets up a notation where "risk" *ex ante* is represented as a probability distribution over sets of probabilities $\{p_i\}$, where each p_i is the probability that the ith individual will die due to the action being evaluated before the end of some time period. He represents "risk" *ex post* as a set of status indicators $\{x_i\}$, where $x_i = 1$ if the ith individual has died due to the evaluated action in the time period and $x_i = 0$ otherwise. He goes on to postulate von Neumann-Morgenstern utility functions U_R on $\{p_i\}$, U_F on $\{x_i\}$, and u on x, where

$$x = \sum_i x_i = \text{the total number of fatalities.}$$

He defines a desire for equity as a preference for more equal p_is, sets up a consistency condition between U_R and U_F, and shows that these conditions lead to his notation to a risk evaluation function as simply $u(x)$ and even shows that it must be convex downward, representing preferences that would prefer a technology characterized by a low-probability-high-loss over one with high-probability-low-loss and the same expected loss. This seems at odds with the general aversion to catastrophe that seems to be found in some public attitudes toward risk. Keeney goes on to add another condition that is consistent with a slightly more complex functional form. While Keeney's evaluation functions incorporate a particular form of equity preference and an attitude toward uncertainty, the form of the evaluation model follows from more of a prescriptive than descriptive set of assumptions.

Bodily (1980) develops another sort of utility function that accounts for several more descriptive aspects than does Keeney's models. Bodily's model accounts for the number of people sharing the risk, the number of casualties per incident, the initial and final states of health involved, changes in individual probabilities of harm or benefit, how voluntary the risk is, and whether lives are being saved or lost. In a set of different examples of risk management alternatives, his calculations show the effective value of life varying by 170 percent due to differences in factors just listed. The key to Bodily's model is the combination of summed willingness to pay to avoid individual risk and a utility measure of attitude toward group risks in a single measure of social risk. While Bodily's model captures an admirable number of factors, it does not involve some other factors that are important, especially equity considerations. That is more or less an artifact of his particular presentation, however. The model framework that Bodily proposes could be extended to include other factors of risk.

The evaluation models proposed by Bodily and Keeney represent promising attempts to model preferences concerning social risk that are not simply linear in numbers of health effects. Their models are basically applications of multi-attribute utility theory, with the value elicitation questions effectively designed by the model. That is, the model is developed, with some consideration for what is important, then the elicitation questions are derived by what is necessary to derive the values of the parameters used in the model. This process leads to elicitation questions that are very relevant to the issues involved, but that are quite unusual and do not occur to people in their normal experience. Approaches for improving this situation are discussed in the next section.

3.2. Psychological Considerations

Research on risk evaluation does not always entail a quantitative model that yields a scalar index of risk. In this section we will discuss psychological research that simply identifies and lists the considerations to be kept in mind in evaluating risk. The research selected as an example is the work of Slovic, Fischhoss, and Lichtenstein (Slovic *et al.* 1980, Slovic and Fischhoff 1981, Fischhoff *et al.* 1978, 1980, Lichtenstein *et al.* 1978). They draw on a large body of experimental work of their own and others to enumerate the ways in which people are bad at probabilistic thinking, including the several biases and heuristics people have been found to use in choosing among alternatives with uncertain outcomes. They then develop implications of those biases and heuristics for problems in managing social risk. As they have done much of their work in problems with acceptance of nuclear power, they list the qualitative aspects of that technology that help to explain the lack of acceptance, as derived from experimental work. Those aspects include lack of control over the technology, dread, lethality of effects, potential for catastrophe, potential for unknown effects, and novelty of the technology. Each of these aspects is defined more precisely in their papers. It should be clear from this list that the work of Slovic, Fischhoff, and Lichtenstein was a primary source for the list of aspects of the psychological perspective on risk presented in this paper. They go on to list problems in evaluating social risk and recommendations for paths to acceptance of technological risk. They do a remarkable job of listing considerations to be kept in mind in evaluating and managing social risk, but do not develop any equations for an evaluation index for risk. In the broadest sense, it could be maintained that they do present a model of social preference concerning risk, in that their lists of biases and considerations explain observed phenomena of societal acceptance and rejection of certain risks. It could also be maintained that their work results in decision aids, in that they list recommendations for decision makers involved in managing social risk. However, they do not develop an evaluation model, if such a model is defined in the narrow sense as a set of equations resulting in a single evaluation index.

3.3. Limitations of the Approaches

The multi-attribute utility approach does not provide the actual utility function to be used, but only the parametrized form of the function along with a protocol of questions to ask somebody whose answers can be used to calculate those parameter values. In other words, the utility function approach does not actually provide any answers, but only rephrases the questions into simpler, more understandable forms that are nevertheless just as difficult to answer. But this approach has even greater difficulties in that there is seldom such a consensus on value trade-offs that it does not matter from whom you elicit the values. That gives rise to a whole host of problems concerning whose values to elicit and how to aggregate different values into a single risk evaluation model (see Arrow 1977 for a review of social choice theory). The second approach discussed, labeled psychological considerations is the most sensitive to the social value aspects of risk evaluation. However, because it does not provide a risk evaluation index, it is not directly useful as a decision aid.

There are two limitations shared by both of the approaches discussed. The first of these is that equity considerations may simply not be appropriate incorporated into any evaluation approach. Such aspects may only be approachable within the bargaining procedure of the political decision making process. An evaluation approach may hope to aid one or more parties in that bargaining, but any attempts to replace that bargaining with a model could be considered inappropriate.

The second shared limitation is perhaps the most serious. That is that the institutional structures making the risk management decision may not be compatible with the decision structure assumed by the evaluation approach. For example, the most basic assumption made by the utility approach is that there is some single self-aware process somewhere where the risks and benefits of the actual decision alternatives are compared. In fact, it is often the case that a regulatory agency, such as the Nuclear Regulatory Commission (NRC), is faced with what appear to be "yes/no" decisions on a single alternative that involves social risk. Of course, the NRC is actually participating in an unorganized way with several other government agencies and private companies in a set of actions which results in a selection of one of the risky alternatives. But where in that set of actions is the single self-aware process that weighs the risks and benefits of the different alternatives? Whose value assumptions and trade-offs should be used in the risk evaluation? Where should the results of the risk evaluation be delivered? It is entirely possible that the biggest limitation in the usefulness of a risk evaluation is the decision making process it is intended to serve (Kunreuther 1980). It could be that any improvement to be made in risk evaluation models would not be as important as it would be to organize the decision making process so that the risks of the actual decision alternatives are compared directly. That type of organization will be referred to in this paper as *decision focused*. Yet a development of a risk evaluation model that is demonstrably sensitive to the aspects of risk management decisions that the political process is sensitive to could go a long way toward encouraging the political process to adopt that desirable decision-focused organization. The development of a risk evaluation model that could achieve that end would be an ambitious undertaking. The next section explores the initial considerations to be kept in mind as one embarks on such a venture.

4. TOWARD A MORE BROAD APPROACH

4.1. General Perspective

The previous sections have discussed research resulting in catalogs of social and psychological considerations, and contrasted that with multi-attribute utility risk evaluation models, which provide a risk evaluation index. The previous sections have also explained the need for a risk evaluation model that lies between the index-producing models and the catalog approach. Such a model would provide a risk evaluation index, and so be more directly useful a decision aid than the catalog approach, yet it would also be more sensitive to societal concerns regarding risk than the index-producing models reviewed above. This section will outline the fundamental considerations necessary to the development of such a model.

The discussion in this section will be based upon two major themes. The first is that a risk evaluation model should not be limited to providing only one of many inputs into a decision making process, but should be a direct aid to that process. That is, the model should not leave entirely to the decision makers the difficult job of putting together a large set of seemingly incommensurate pieces of data. Rather, it should deliver results in a form that aids the members of the process to integrate those results with the other considerations that enter in. The model should be designed to elevate the level of debate by providing a structured framework of reasoned evaluation within which the decision can be made.

The second theme which forms the basis of the following discussion is that the risk evaluation model should be built around risk considerations that drive the political decision making process. The order of development of the model should be from the primary concerns of the process to the value elicitations that would capture those concerns to the form of the model that would be logically fitted to those elicitations. This is a subtle theme, at best, and one not incompatible with the models of Keeney and Bodily. The basic idea is to start from the observables that people react to in a risk management problem, and build the model around those.

4.2. General Methodology

The general methodology adopted here is based on multi-attribute utility theory. That discipline is only briefly described here. The reader is referred to a basic text for a complete description (e.g., Keeney and Raiffa 1976). The methodology is based on the fitting of a multi-attribute utility function (i.e., a multi-argument von Neumann-Morgenstern utility function), to the answers a person gives to a structured set of questions. The basic aspects of preference captured by the function are trade-offs between conflicting objectives (i.e., slopes of indifference curves) and attitudes toward uncertainty. The key development of the methodology is a set of theorems relating various plausible and testable assumptions about preferences to corresponding simple forms of multi-attribute utility functions (MAUF). The fitting of the MAUF in a value elicitation interview then amounts to a set of assumption tests to identify the most simple acceptable form of the function, followed by the set of judgments necessary to fit the parameters of that functional form. Those judgments include indifference map comparisons and preference comparisons of simple uncertain alternatives. The fundamental concept underlying the methodology is that individuals are not good at choices between complex alternatives (such as actions involving multidimensional and uncertain outcomes), but can make choices between simple alternatives in a manner consistent with their underlying preferences. The methodology provides a mathematical system which can take alternatives too complex for consistent intuitive judgment, and evaluate them in a manner consistent with a person's judgments concerning simple alternatives, where his judgments are more apt to reflect accurately his underlying preferences.

In typical expositions of the decision analysis methodology, the first steps in building the evaluation function are to identify the objectives of the decision maker, and describe an observable measure for each objective that represents the degree to which each alternative satisfies that objective. This section identifies typical objectives of the decision making process concerned with social risk management, taking the psychological perspective of risk described earlier. Each objective is identified and characterized in a way that suggests possible observable measures that could be used to develop the evaluation function itself and its corresponding elicitation procedure. As this section represents only a first effort at the development of a comprehensive risk evaluation model; further steps in model development are left to future research.

4.3. Objectives of Risk Management from a Psychological Perspective

In the objectives listed below, it is assumed that some single-dimensional measure of social cost has been decided upon that evaluates all the heatlth and environmental effects upon an individual in some standard unit. There is no intention to suggest that such a measure would be easily derived, as it would require answers to such questions as: How many colds is a cancer worth? However, as the objectives listed involve considerations above and beyond the measure of the severity of a deleterious effect upon an individual, it clarifies the discussion considerably to assume that such a measure exists.

As a second concession to clarity, the following discussions focus on the evaluation model itself, and so do not consider such frequently dominant problems as disagreements on the probabilities to be used in calcualting the evaluation measure.

4.3.1. Expected Number of Health Effects

While the expected number of health effects is a common measure of risk from a technological perspective, it can also be considered relevant to the psychological perspective. It is most relevant when the health effects are lives lost or saved, and expected lives lost or saved are considered equivalent to actual lives lost or saved. In that sense the expected fatalities measure has an ethical basis that leads to a troubling dilemma concerning the addition of other objectives to an evaluation function. A risk management strategy using simply expected fatalities as an evaluation function will, if successful, minimize expected lives lost. Any strategy using a more comprehensive evaluation function will allow more expected lives lost in any nondegenerate case. Thus that comprehensiveness in evaluation has "cost" some increment in expected lives lost (or decrement in expected lives saved). While that increment would not actually be a cost in terms of the social welfare represented by the evaluation function, it would nevertheless be the case that any attempt to use a more comprehensive evaluation model would be open to the attack that its use would cost human lives, expectationally.

This first objective is exceptional in that the measure of the objective to be used in the evaluation model is the objective itself. The major measurement problems involved in this objective are hidden in the clarifying conventions presented at the beginning of Section 4.3. The remaining objectives listed below do not have such obvious measures associated with them.

4.3.2. Possibility of Catastrophe

While the idea of possibility of catastrophe may be intuitively clear, closer examination reveals a complex concept. As has often been mentioned in comparisons of coal and nuclear power as sources of electricity, while expected fatalities per megawatt-year (electric) may be estimated as less for nuclear than for coal, the fact that the nuclear fatalities occur in high-fatality, low-probability accidents explains public resistance to nuclear power (Barrager *et al.* 1976). This comparison will be mentioned again below, since in fact there are a number of reasons that could explain that resistance apart from the possibility of catastrophe. However, for this discussion the example is useful in considering just what catastrophe means. It seems clear that a catastrophe is a "bunching" together of fatalities, but on what dimension does that bunching occur? Fatalities can be bunched together in one place, or at one time, or in one state of nature, or by one identified cause, or in one previously identifiable group of people. These various types of bunching may correspond to different types of preference mechanisms.

The risk of a nuclear reactor involves bunching on all five of the dimensions listed: fatalities are concentrated in states of nature that correspond to an accident, are grouped downwind of the facility at the time of the accident, are bunched by single cause, and impact the neighbors of the reactor. A coal plant induces fatalities that occur with some degree of certainty and so are distributed over many states of nature, and are distributed in space, time, and groups of people. Coal plant fatalities are not even bunched by identifiable cause, in that respiratory ailments caused by the coal plant cannot be distinguished from ailments from other causes. For example, fatalities due to nuclear waste may be bunched by state of nature (failure of casings, misunderstanding of physical processes), and bunched by location, but not bunched by identifiable cause, and distributed over very long periods of time and groups of people (generations). Risk due to train derailments of toxic chemicals may be concentrated in the poor segments of the population who live near railroad tracks, although it is distributed over several accidents (in different states of nature, places, times), each with very few fatalities.

Bunching by identifiable cause is evaluated in the attributability objective discussed later. Bunching by identifiable group is evaluated as part of the concept of equity also discussed below. That leaves catastrophe defined here as a bunching in space, time, or state of nature. Another way of defining a catastrophe is as a number of fatalities linked in any way. By either definition, the problem remains to evaluate attitude toward catastrophe. The measure could be a set of probability distributions over numbers of health effects in each different type of catastrophe, if more than one type is possible.

Efforts thus far in the evaluation of potential for catastrophe have taken three paths. First, a nonlinear value function over number of fatalities has been suggested, either as an *ad hoc* function or as a von Neumann-Morgenstern utility function (Keeney 1980a). Keeney speculates that such a curve would be convex downward due to riskless preference effects (diminishing marginal disutility for health effects). That convexity represents a preference for bunched over distributed fatalities. Second, a multi-attribute utility function has been suggested, where the health status of each person is an attribute (Bodily 1980). With such a function, the interaction terms represent attitudes toward catastrophe. With both of these evaluation strategies, if only the suggested utility function is used the evaluation is only sensitive to bunching in states of nature, and does not account for differences in bunching in space and time. Neither model has the required argument structure or elicitation protocol to be sensitive to bunching in space and time that is not bunched by state of nature. While the required extension of the model's notation would be relatively straightforward, attempts to extend the elicitation protocols as necessary reveal some fundamental problems to be discussed in a subsequent paper.

4.3.3. *Equity*

Just as discussions concerning risk often suffer from lack of a definition of risk, discussions concerning equity may involve as many as three distinctly different concepts going by the same name. Each concept is discussed in turn here.

(a) *Correspondence between populations at risk and at benefit.* Strip mining may involve risk externalities imposed on people of the mining region so that people in a distant metropolis can run air conditioners. This form of equity problem has been addressed at great length in the social choice literature, and so will not be discussed at any depth here. It should be noted, however, that no consensus exists as to the best method for evaluating a situation that is inequitable in this sense (Arrow 1977). After risk evaluations for each of the populations whose members are roughly equally affected, the use of those measures in a decision may be one aspect of risk evaluation best left to the political process.

(b) *Ease of identifying people at risk: size of individual probabilities.* While the size of an individual's probability for a health effect is of interest for willingness-to-pay calculations, in this discussion that number is used to represent a different concern: how much society knows about who and how many will be victims. Suppose a technology is estimated to cause one expected fatality (EF). The size of the population bearing that risk would be of great importance to its evaluation. Consider that 1 EF risk to one person is very different from 1 EF risk equally shared by 100 people, or 1 EF risk equally shared by 10^8 people. The difference to society between these cases is in its level of knowledge as to *who* and *how many* may die. At a fixed number of expected fatalities, the variance over numbers of fatalities rises with the number of people equally sharing the risk, though only very slightly once there are more than 20 people. While this sensitivity to the size of the group sharing one EF of risk could be confounded with sensitivity toward attributability, discussed below, it could not be confounded with sensitivity toward catastrophe, as there is no consideration for intercorrelation of the occurence of the fatalities.

The relationship between differences in individual probabilities and variance in number of fatalities is that for a given level of expected fatalities, the less different the individual probabilities, the greater the variance in number of fatalities. This relationship is interesting in that it explains a problem presented by Keeney (1980b). Keeney establishes an evaluation of numbers of fatalities by a von Neumann-Morgenstern utility function over that number. He points out that an aversion to catastrophe would correspond to a concave downward, or uncertainty averse, utility. He then defines a desire for equity as a desire to make any two individual probabilities of dying less different. Finally, he shows that such a desire for equity corresponds to a convex downward utility function, and so is incompatible, in his model, with aversion to catastrophe. The significance of this conclusion becomes more clear when one realizes that any change to increase equity (his definition) also increases the variance in number of fatalities. That is, Keeney has basically made the statement that "equity is uncertainty". It is from that statement that his conclusion flows most directly. This matter is an excellent example of how surprising conclusions can be drawn from simple evaluation models and simple definitions of measures of social concerns, such as equity and potential for catastrophe.

(c) *Ease of identifying people at risk: ease of defining the impacted group.* As with the previous concept, the key to this concept of equity is knowledge, but in this case in a more subtle way. Suppose some evidence in science enables the certain identification of a victim before he is killed by a technology. Even though society now has perfect knowledge as to who the victim will be, there may not be a perception of inequity if the victim cannot be easily described. If there is nothing in common among the victims except that they are American, say, then the impact of the technology may be seen as equitable. If, however, all victims are black, or all asbestos workers, or all live downwind of a nuclear reactor, or all are poor, the impacts would be seen as inequitable, regardless of how large or small the group sharing the risk is.

The three general considerations described so far (expected number of effects, catastrophe, and equity) are three distinct measures of social risk. A three-dimensional table could be laid out, with each cell filled with an example representing a different combination of levels of the three measures of social risk. Each of those examples could be described by a probability distribution over various dimensions, such as numbers of health effects, classes of people, space, and time. The remaining considerations, discussed below, are not so amenable to quantification.

4.3.4. Degree of Control

The concept of degree of control has two different aspects: an impacted individual's participation in the decision to expose himself to the risk, and an individual's participation in the decision to deploy the technology. The former aspect coincides with the voluntary *versus* involuntary risk distinction made by Starr (1969) and discussed by others (Otway and Cohen 1975). Starr presented evidence which he interpreted as indicating that society has a much higher threshold of acceptability for risks involving voluntary exposure than for risks incurred involuntarily. Arguments against that hypothesis have appeared in the literature, most recently in the paper by Slovic and Fischhoff (1981), which concludes that apparent aversion to involuntary risk can be better explained by the higher potential for catastrophe and inequity that often accompany that type of risk. However, another paper involving the same two authors stresses the importance of public participation in the second of the two decisions listed above, concerning the deployment of the technology (Fischhoff *et al.* 1980). That second decision involves an aspect that is basically different from any consideration mentioned so far, in that it has little to do with the physical source or impactee, but addresses the process which generated that source and impactee. As some have pointed out (Green 1981) one determinant of that elusive concept called acceptable risk is the acceptability of the process that generated the risk.

4.3.5. Attributability

Attributability of cause is the first in a series of considerations that involve very large steps away from a model that could be easily defended on prescriptive grounds, steps toward a descriptive model. In some ways, this consideration is related to equity: while some aspects of equity have to do with ease of identifying an impactee, attributability concerns the ease of identifying the source of the effect. Attributability is also highly correlated with catastrophe, while it would be much harder to identify the cause of a more distributed set of effects. If a set of effects is clearly attributable to a single cause, that cause is more apt to be discovered and to give rise to societal corrective mechanisms than is a cause with more subtly distributed effects. A gas tank design feature that makes a car go up like a torch when hit is much more apt to be reacted against than a steering design flaw that leads to crashes erroneously attributed to driver error.

The last example suggests that attributability, like equity, is an aspect very dependent on level of information. As epidemiological studies and national medical reporting systems become more effective, diseases that would have gone unnoticed and unfeared a few years ago become centers of attention and concern. Effects of pesticides, defoliants, and other widely dispersed chemicals cause concern only when they are recognized as coming from controllable chemicals, yet those effects may only be able to be detected by very sophisticated techniques, if at all.

Attributability has more than the two levels: recognized and unrecognized. It concerns how easily the cause can be identified. Of course, a society will only react to risks that are recognized, but beyond that, the ease with which the cause can be described will affect the ease of ideation of the risk as a menace. To return to the car gas tank design example, the image of an engineering staff making a fairly well defined decision not to incorporate a safety feature is a very clear target for public reaction. It could be that a general lack of competence of an automotive engineering staff could contribute much more to driving risk than any single decision about gas tanks, but general lack of competence is much harder to identify, and even if identified, forms a much more diffuse image for public reaction to try tc focus on. While these aspects are part of a descriptive basis for attributability as a dimension of risk, they also suggest a prescriptive basis, in that the ease of identification of a cause is apt to reflect very closely the ease of management of the risk.

4.3.6. Nonprobabilistic Factors

Perhaps the most important dimension of social risk from the psychological perspective is the size of the maximum potential catastrophe, considered *without* any weighting by its probability. The idea of evaluating an alternative by a possible outcome without considering its probability is a very large step away from a prescriptive evaluation model and toward a descriptive one. It can be extremely troublesome for the analyst, not only because it doesn't seem to make sense to someone used to probabilistic models, but also because the description of a maximum potential catastrophe is dependent on the imagination of the analyst, and on an ill defined notion of what "potential" is. Yet the political process typically does not share the probabilistic perspective that is so fundamental to risk assessment. One example involves the attempt to site a liquefied natural gas (LNG) terminal at Oxnard, California. While risk analyses found the site to be safe, maximum credible accident (MCA) scenarios involving very large numbers of fatalities formed a large part of the political debate which led to the effective rejection of the site. That debate did not seem very sensitive to the extremely low probabilities that could be assigned to the MCAs (Ahern 1980).

There are a number of explanations that could be invoked for this nonprobabilistic evaluation. First, it could be that members of the political process do not know how to think probabilistically, or do not have a feel for what a very low probability means. Second, members of the political process may simply doubt the low probabilities. That doubt may be well founded, as it is impossible in any actual case to guarantee the completeness of any risk analysis, and major accidents can be identified that were pronounced impossible before their occurrence (e.g., the *Titanic*), or that were not effectively described by a preceding risk analysis. In a particularly strict sense, no absolute probabilities are ever calculated by a risk analysis. The probabilities reported by such analyses are conditional on the validity of the assumptions of the analysis, including those concerning its completeness and descriptive validity. It may be very reasonable for members of the political process to doubt that an extremely low probability is in fact that low, given its conditional nature.

A third reason for nonprobabilistic evaluation is a desire for resilient social support systems. Given the extreme uncertainty in what future demands will be made on, for example, an energy supply system, and the uncertainty in the behavior of that system, it may make sense to design it to be as resilient as possible. One way to promote resilience is to limit the maximum potential catastrophe or MCA.

A fourth reason for nonprobabilistic evaluation is a sensitivity to the concept of dread as a social cost. If a large plant is erected that has an MCA involving the deaths of 40 000 people living downwind, those people are going to live their lives with a sense of dread that is not very directly related to the very low probability a risk analysis might assign to the MCA. That sense of dread may be a very real decrement in their quality of life, regardless of how safe the plant is as measured by any risk analysis.

It may be helpful to give some meaning to the idea of limiting the MCA of an option with a brief example. The most dramatic MCAs involve accidents at large-scale facilities that send poisons or flammable gases into the air that can cause large numbers of fatalities downwind. Examples include nuclear reactors, LNG terminals, and chemical plants. In any of these cases, the MCA can be limited by requiring that the facility be sited remotely, but only at a financial and perhaps environmental cost.

4.3.7. Nondecision Comparisons

All the facts discussed above could at least be fitted into a decision analytic framework, in that each could be involved in a comparative evaluation of at least two alternatives in a decision. However, there are comparisons made in some risk analyses that have nothing directly to do with the decision being considered (Cohen and Lee 1979). It makes immediate sense to compare the risks of a coal electricity generating plant with a nuclear one, as those are two realistic alternatives considered by a utility in expanding its capacity. It makes much less immediate sense to compare the risks of living near a nuclear reactor with the risk of moving to Denver, driving a car an extra three miles, or smoking an extra pack of cigarettes. The last three comparisons do not relate very directly to any decision, as few people actually do choose between a reactor neighborhood and Denver, or between living near a reactor and smoking more.

The intent of the nondecision comparions listed above is to fit some technological risk into the same scale with risks people normally accept, then go on to draw conclusions regarding the acceptability of the technological risk. There are a number of problems with this approach as a way to establish the acceptability of a risk. However, nondecision comparisons are based on an idea that could be incorporated in a risk evaluation model: people make choices every day that involve risk, and those typical risks may form benchmarks on an individual's scale of perceived risk. There are problems in that the various risks considered are evaluated differently (nuclear power, background radiation, driving, smoking), but once those differences are accounted for within an evaluation function, routine decisions involving risk could be used as meaningful points of reference on the scale of the risk evaluation index (see Lichtenstein *et al.* 1978).

4.3.8. Nontechnological Factors

While the set of factors described so far has spanned a wide range of characteristics, all of them have been concerned with the deleterious health and environmental effects of the evaluated technology. In this final step in the progression of factors, even that one common thread is abandoned as nontechnological factors, factors other than health and environmental effects, are considered. A number of analysts have observed that resistance to nuclear power may be based in part on a resistance to the political side effects of the technology: the centralization of power that might accompany a centralized generation of electricity; the loss of civil liberties that might result from efforts to prevent diversion of nuclear material, etc. There might also be resistance to perceived effects on lifestyle: the more materialistic, growth based, exploitative society made possible by nuclear power; the less natural, less simple, less human society associated with the high technology of nuclear power. While the attribution of such far-ranging effects to nuclear power is certaintly arguable, the fact remains that there is a body of survey evidence establishing that segments of the population do ascribe these effects to nuclear power (Otway and Pahner 1976).

It is one thing to establish that nontechnological effects are important to some people's evaluation of a technological alternative. It is quite another problem to decide how that fact would be used in developing a risk evaluation model. The incorporation of nontechnological effects would be required in a risk evaluation model intended to predict political resistance to the deployment of a technology.

Yet that incorporation would expand such a model beyond the role of evaluating risk. It seems that the decision whether or not to attempt to include nontechnological effects in the model is a decision on model scope. Is the model to be a risk evaluation model or is it to be a resistance prediction model? The discussions in this paper are aimed at the development of a risk evaluation model sensitive to societal concerns. That model does not have to be a resistance prediction model in order to aid the political decision making process in evaluating social risk. It follows, then, that while nontechnological factors are a useful end point in a progression of social concerns, they are not necessarily to be included in a risk evaluation model. Perhaps the most useful role such factors play in the discussions of this paper is in defining the limits to which a risk evaluation model should go in the progression from technical risk evaluation to aiding the political decision making process.

4.4. Incorporating Risk Evaluation Factors into a Risk Evaluation Function

The previous section listed eight factors involved in the evaluation of social risk from a psychological perspective. While it has been argued that the eighth factor is not appropriate for inclusion in a risk evaluation model, the other seven are. The development of a risk evaluation model is then defined as the development of measures of each of the seven factors, then the combination of those measures into a single index. While these two tasks are the subject of future research, this section briefly comments on some aspects of the two tasks apparent from an examination of current risk evaluation work.

The first thing to note about the problem of incorporating the different factors into a single evaluation model is that each of the existing measures incorporates only a few of the various factors. Rasmussen curves, for example, measure the potential for catastrophe, could be used to define a maximum credible accident as a nonprobabilistic measure, and have been used for nondecision comparisons of social risk (e.g., reactor risk *versus* meteorite risk). However, such curves do not reflect any of the other factors. Descriptions of risk involving the set of individual probabilities of health effects $\{p_i\}$, lend themselves to reflecting concerns about equity, and are appropriate for nondecision comparisons at an individual level, but do not reflect the other factors. Of course, Rasmussen curves could be calculated for each of several subgroups of the population to address concerns about equity, and the work of Keeney (1980b) and Bodily (1980) extends evaluations of $\{p_i\}$ to include concerns about catastrophe, but the fact remains that no existing evaluation measure or model addresses all seven of the factors described.

Sometimes the assumptions involved in a risk evaluation model can restrict in a very subtle way which factors are addressed. For example, Keeney (1980b) sets up a notation and a basic model in such a way that assumptions about treating everyone equally and correspondence between *ex ante* and *ex post* utilities lead to a single-attribute utility function on number of fatalities as a risk evaluation measure. While his notation, basic model, and assumptions are intuitively compelling, the resulting measure is sensitive only to expected numbers of fatalities, to bunching by state of nature and to equity in an equally shared p_i sense. The other factors are missed not because of obvious features of the model, but because of a lack of dimensionality in the original situation to be evaluated.

The example of Keeney's model is given because it illustrates the importance of the first step in developing a comprehensive risk evaluation model: the description of the effect space to be evaluated. Keeney's description of the effect space as a probability distribution over $\{p_i\}$ accounts in large part for the nature of his resulting risk evaluation measure. The same information cast into a catastrophe-by-equity-by-expected fatality space could lead to a very different risk evaluation measure, but one still limited in the factors it could address. In order to address all factors, an evaluation model must begin with an effect space in which all factors are represented. Because the seven factors involve such different aspects as bunching of effects and the participatory nature of the process, there are several different ways to set up the effect space. Clearly, the first step in future research on the development of risk evaluation models is an examination of those different effect spaces to select the one most appropriate for a meaningful value representation and elicitation procedure.

5. SUMMARY

This paper has discussed the problem of developing better risk evaluation models. After an examination of the central problem of risk evaluation, defining what risk is, two risk evaluation approaches were presented that bracket a particular part of the analysis spectrum in need of development. That approach involves the expansion of multi-attribute utility risk evaluation functions to be more sensitive to social and political concerns regarding risk management. Initial considerations for research in that direction were laid down. That approach appears to be feasible, and promises a more useful role for analyses in the process of social risk management.

The most unusual feature of the broad approach proposed in this paper is that it seeks to recognize explicitly some of the political realities involved in the social risk management process. An attempt is proposed to develop an evaluation model that deviates from prescriptive considerations to account for preferences expressed in the political decision making process. This attempt is marked by several difficulties, perhaps the most notable among them being the ethically troublesome result mentioned earlier: evaluation models sensitive to any feature in addition to or other than expected fatalities will generally advise the selection of policies which do not maximize expected number of lives saved. Policy makers using such models could be accused of "sacrificing lives, expectationally" for the sake of being responsive to other societal concerns. Of course, expected lives are allowed to be lost to gain on other dimensions all the time in public policy decisions, but it may be difficult to do so deliberately, as an explicit step in the policy making process.

The main potential benefit of the proposed broad evaluation model is that it seeks to avoid the problem of a technically correct risk analysis advising the promotion of a project that is eventually rejected due to popular opposition. Another benefit is its general decision aid orientation, an improvement over standard risk analyses which are intended only as inputs into the decision making process.

There are two major limitations to the broad approach proposed. First, the more explicit use of subjective judgments involved in the value elicitations can cause problems with the defensibility of the approach. This is in turn a problem with the political decision making process, since those subjective judgments generally must be made somewhere in the decision making. They are either hidden in the assumptions of the model, or they are explicit. But the fact remains that a government agency may have more trouble defending a decision based on an analysis with explicitly subjective judgments than one based on an apparently objective analysis.

The second problem is more basic, but less specific to the broader approach proposed. That problem is that better evaluation models may not help social risk management as much as would better consideration of actual decision alternatives. As was mentioned in an earlier section, risk analyses are typically not part of a political decision making process that recognizes the choice between several risky alternatives. For example, the Nuclear Regulatory Commission seems to be faced continually with decisions that have the appearance of being "yes/no" decisions on a risky option, when it is actually participating in an unorganized way with several other government agencies and private companies in a process that makes "which" decisions, choosing one of several risky alternatives. A political process that explicitly faced such "which" decisions would be able to make better use of comparative risk analyses (Ahern 1980). But perhaps this identifies another potential advantage of the proposed approach. A risk evaluation model that is oriented toward dimensions of concern to the political process could be designed in such a way that comparison with the actual alternative is an intrinsic part of the evaluation.

REFERENCES

Ahern, W. (1980) California meets the LNG terminal. Coastal Zone Management Journal 7:185-221.

Arrow, K. (1977) Current developments in the theory of social choice. Social Research 44(4).

Barrager, S.M., Judd, B.R., and North, D.W. (1976) The Economic and Social Costs of Coal and Nuclear Electric Generation. Washington, D.C.: National Science Foundation.

Bodily, S.E. (1980) Analysis of risk to life and limb. Operations Research 28(1).

Cohen, B., and Lee, I. (1979) A catalog of risks. Health Physics.

Fischhoff, B., Lichtenstein, S., Slovic, P., Keeney, R., and Derby, S. (1980) Approaches to Acceptable Risk: A Critical Survey. NUREG/CR/1614. Washington, DC: US Nuclear Regulatory Commission.

Fischhoff, B., Slovic, P., Lichtenstein, S., Read, S., and Combs, B. (1978) How safe is safe enough? A psychometric study. Policy Sciences 8:127-52.

Green H. (1980) Implications of nuclear accident preparedness for broader nuclear policy, in Planning for Rare Events: Nuclear Accident Preparedness and Management, edited by J. Lathrop. Oxford: Pergamon.

Hazelwood, R.N., and Philipson, L.L. (1977) Survey of LNG Risk Assessment. Report for the California Public Utilities Commission. Los Angeles: Socio-Economic Systems, Inc.

Keeney, R.L. (1980a) Evaluating alternatives involving potential fatalities. Operations Research 28(1).

Keeney, R.L. (1980b) Equity and public risk. Operations Research 28(3).

Keeney, R.L., and Raiffa, H. (1976) Decisions with Multiple Objectives. New York: Wiley.

Kunreuther, H. (1980) Societal Decision Making for Low Probability Events: Descriptive and Prescriptive Aspects. WP-80-164. Laxenburg, Austria: International Institute for Applied Systems Analysis.

Lichtenstein, S., Slovic, P., Fischhoff, B., Layman, M., and Combs B. (1978) Judged frequency of lethal events. Journal of Experimental Psychology: Human Learning and Memory 4(6).

Linnerooth, J. (1978) Reevaluating the value of life: Practical considerations. Invited paper presented at ORSA/TIMS Conference, Los Angeles, November.

Otway, H.J., and Cohen, J.J. (1975) Revealed Preferences: Comments on the Starr Benefit-Risk Relationships. RM-75-5. Laxenburg, Austria: International Institute for Applied Systems Analysis.

Otway, H.J., and Pahner, P.D. (1976) Risk Assessment. Futures 8(2).

Pate, M.-E (1978) Public Policy in Earthquake Effects Migration. Technical Report 30. Stanford: Blume Earthquake Engineering Center.

Rothschild, Lord (1979) Risk. Atom 268:February.

Slovic, P., and Fischhoff, B. (1981) How safe is safe enough? Determinants of perceived and acceptable risk, in Too Hot To Handle: Social and Policy Issues in the Management of Radioactive Wastes, edited by L. Gould and C.A. Walker. New Haven: Yale University Press.

Slovic, P., Fischhoff, B., and Lichtenstein, S. (1980) Facts and fears: Understanding perceived risk, in Societal Risk Assessment: How Safe is Safe Enough? edited by R.C. Schwing and W.A. Albers. New York: Plenum.

Starr, C. (1969) Social benefit versus technological risk. Science 165:1232-8.

UNSNRC (1975) Reactor Safety Study (Rasmussen Report) WASH-1400. Washington, DC: US Nuclear Regulatory Commission.

Group Decision Making Methods for Evaluating Social and Technological Risks

Paul R. Kleindorfer

Decision Sciences Department, The Wharton School, University of Pennsylvania, USA

1. INTRODUCTION

This paper discusses group choice processes in evaluating social and technological risks. We first outline the rationale for government involvement in regulating certain classes of risks. This discussion highlights the social roots of policy making in the risk area. In particular, if one takes the normative perspective of trying to make policy choices in the name of certain stakeholders (e.g., the public at large), then the problem naturally arises of identifying and incorporating the possibly conflicting values and perceptions of such stakeholders into evaluation and choice procedures. Direct participation (e.g., through referenda) is one mechanism for accomplishing this. Other mechanisms include public opinion surveys and the use of panels and committees composed of community leaders or other role players representing the interests of stakeholders. These latter mechanisms, as well as the policy making process itself, lead to a discussion of controlled group processes as a means of accomplishing certain evaluation and information gathering tasks in the policy formation process. After structuring various forms of group activity (e.g., forecasting, scenario generation, and choice), an overview of recent research on group processes as it relates to these issues is presented.

2. REGULATING RISKS AS A PUBLIC POLICY PROBLEM

Government involvement in assessing and regulating social and technological risks has been justified on the grounds that market based solutions will probably fail to provide either efficient or equitable insurance against relief from the consequences of such risks (see Lowe 1972, Kunreuther *et al.* 1978, Zerbe and Urban 1980). The reasons for such market failure are classic. First, many risks entail involuntary exposure on a broad scale, and thus, individuals may find it impossible to take effective preventive precautions or, alternatively, they may not be able to appropriate all of the benefits of such protective activity to themselves. These excludability problems allow one to classify many social and technological risks as public "bads" for which market solutions have well known deliberating features (see Arrow 1970). Secondly even if the liabilities for and consequences of such events could be identified and meted out in some acceptable manner *ex post*, the costs for individuals to obtain and process the information necessary for them to make rational protective and avoidance decisions *ex ante* may be prohibitive. In addition to these problems, the magnitude of the financial risks involved may make the government the only reliable insurer.

These possible reasons for market failure suggest that several basic dimensions of risks are important in determining when the benefits of governmental regulation of risks will exceed the costs of such intervention. First, the nonexcludability (or involuntary) nature of the risk; secondly, the magnitude of possible losses involved; and third, the complexity of the evaluation task in assessing the risk and choosing appropriate protective measures*. When these three factors are all strongly

*See Fischhoff *et al.* (1980), Starr and Whipple (1980), and Kleindorfer and Kunreuther (1981) for a discussion of these dimensions indicating their relevance in assessing

evident, as in new drugs and nuclear power, one has a *prime facie* case of governmental regulation of the risk in question. Unfortunately, as we have learned, these same factors lead to severe difficulties in the regulatory process.

To clarify some of the difficulties involved, let us begin with a simple but formal statement of the problem which might confront a public agency charged with designing and evaluating policy options for dealing with a given risk*. Suppose that the agency is concerned with a set of welfare measures $W = (W_1,...,W_k)$. These measures would typically be indicators of the economic, social, and public health and safety consequences of agency activities and policies.

Let us denote the stakeholders in this analysis by the set $N = \{1,2,...,n\}$. (We assume the agency knows who they are and how to classify them.) These stakeholders would be the various segments of the public, individually and as organizational entities, whose values and welfare are deemed important for policy purposes. We shall consider each stakeholder group to be homogeneous, so that we will speak of the preferences, behavior, etc., of stakeholder (group) i. Let us suppose that the welfare (or felicity) of the ith stakeholder group may be represented through a set of indices or attributes (such as economic, psychological, and risk exposure indicators), which we denote by $A_i = (A_{1i},...,A_{mi})$. Thus, we will assume that the attributes A_i are sufficient for policy purposes to evaluate the impact of policy alternatives on group i. The relationship between stakeholder welfare (A_i) and agency objectives (W) will be clarified below. For the moment we note that we generally have in mind a social utilitarian paradigm for which the agency's welfare indicators depend on policy options through their implied consequences for stakeholder welfare, i.e., $W = W(A_1,A_2,...,A_n)$.

The welfare indicators W and $\{A_i | i \in N\}$ may be expected to depend on variables such as stakeholder wealth, goods bought and sold in the economy, the policy option chosen by the agency, and the output of the (risky) regulated sector as measured by cost and risk exposure indicators. In addition, one would expect such welfare indicators to depend on a set of expectations (of "relevant states of the world") over the planning horizon of the stakeholders. In the terminology of strategic planning, we may characterize all of these variables collectively as a "scenario". Such scenarios will condition the preferences and behavior of group i and different groups may have different preferred scenarios because of differences in values and in expectations.

The relationships between the quantities defined above are shown in Figure 1. The agency chooses policy options in the face of political and social processes which determine controls on the agency and filter information on stakeholder welfare. The policy choices have two effects. First, they affect the well being of stakeholders and therefore they may also affect stakeholder behavior in response to the policies chosen. Secondly, of course, the policy chosen may affect the output of the risky sector.

This discussion suggests the following procedure as a guide to policy analysis**. First, the agency predicts technological consequences and stakeholder responses to each of the policy options (the *descriptive phase* of policy analysis). The results of this descriptive phase are the primary input to the *prescriptive phase* of the analysis, where conflicts of values and facts must be resolved in reaching a policy choice. In slightly more detail, the design components of prescriptive policy analysis are the following:

(1) the operational specification of agency goals W and their relationship to stakeholder welfare $\{A_{ij}\}$ and technology;

---

political acceptability and the probable benefits of governmental regulation. See also Lawless (1977) for interesting case studies corroborating the importance of these dimensions.

*Following the seminal work done at IIASA and IAEA (see e.g., Greer-Wooten 1980), risk analysis may be thought of as consisting of risk identification, risk evaluation, policy choice, and risk management. We are concerned here with evaluation of already identified risks.

**See Lasswell (1971), Luft (1976) and Kleindorfer and Kunreuther (1981) for an expanded discussion.

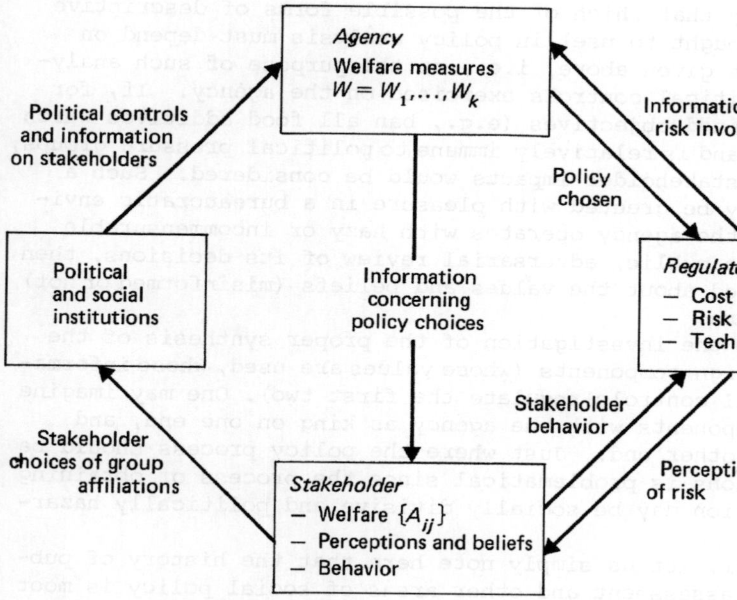

Figure 1. Illustraing policy analysis for risk.

(2) the process by which W, A_{ij}, and technology are measured and evaluated in the descriptive phase;
(3) the process by which agency policy is reviewed and legitimated.

The problem of linking prescriptive and descriptive anlaysis should now be clear. In very simple terms, what one can properly prescribe and defend depends on what underlying behavioral and technological assumptions one is willing to make in evaluating the consequences of various policy options. The sources of conflict in evaluating technological risks are therefore of two types: conflicts of value, and conflicts of "fact". Concerning conflicts of values, these can be irremediable or they may be removable by stilling the woes of injured parties through monetary transfer payments (e.g., O'Hare 1977). Concerning the facts, the question is: Whose scenario will be used as a backdrop against which policy options will be evaluated? Should these be the scenarios entertained by group i (representing its *ex ante* perceptions and beliefs about uncertain states of the world) or should these be somehow "corrected" to represent the possibly better information available to the policy maker? These are issues about which we will have more to say shortly. For the moment, let us summarize the above discussion by noting that there may be many possible (i.e., defensible) descriptive models delineating the impacts of policy choice. Such differences may arise from*:

(i) errors and imprecision in defining the indicators A_{ij};
(ii) differences in assumptions about how to describe the relevant future against which to evaluate policy alternatives (these differences may have deep social and cultural roots; see, e.g., Wynne 1980, Thompson 1980);
(iii) differences in assumptions about the technological consequences of policy;
(iv) differences in assumptions about the decision processes and resulting (predicted) behavior of stakeholders.

---

*Indeed, one may argue that the major accomplishment of the recent literature on risk assessment has been to document the tremendous variability in values, perceptions, and decision processes of the public in response to technological risks and natural hazards. See e.g., Conrad (1980), Fiddle (1980), Fischhoff *et al.* (1980), Kunreuther *et al.* (1978).

Now one suspects immediately that which of the possible forms of descriptive models our agency would use (or ought to use) in policy analysis must depend on the prescriptive design variables given above, i.e., on the purpose of such analysis and on the nature of the political controls exercised on the agency. If, for example, the agency has well defined objectives (e.g., ban all food additives which fail an easy to administer test) and is relatively immune to political pressure groups, then only the agency's views on stakeholder impacts would be considered. Such a "clean" situation would naturally be greeted with pleasure in a bureaucratic environment. On the other hand, if the agency operates with hazy or incommensurable objectives, and is subjected to a public, adversarial review of its decisions, then it would (and should) be concerned about the values and beliefs (misinformed or not) of political powerful stakeholders.

This discussion leads us to the investigation of the proper synthesis of the above mentioned prescriptive design components (whose values are used, whose information is used, and which political controls regulate the first two). One may imagine a spectrum for each of these components with the agency as king on one end, and various stakeholders ruling the other end. Just where the policy process should be located on each of these dimensions is problematical since the process of obtaining stakeholder values and information may be socially divisive and politically hazardous for agency officials.

To develop this idea further, let us simply note here that the history of public participation in technology assessment and other areas of social policy is moot on how best to organize such participation for prescriptive purposes*. The problems with predicting the benefits of participation have been legion. Simply put, the areas where one would most like to have a direct and interactive assessment of public perceptions and values through participation are usually those plagued by the complexity and uncertainty which confound consensus and lead to conflict**.

Against this background, what I propose to do in the remainder of this paper is to review some results of controlled group research on various well defined group tasks. My rationale in undertaking this is to highlight the delicate balance which group researchers have found to exist between effectiveness and group process and thereby shed some light on the general question of participative mechanisms for problem solving. My second reason is that one direct way of accounting for stakeholder values and perceptions is through controlled group processes directed at role playing of various stakeholder groups.

3. GROUP DECISION PROCESSES AND PLANNING

Throughout this paper we will assume that a group (of either stakeholders or policy makers) has been assembled in response to some problem situation. One may discern three major research traditions that would be helpful in understanding the problems confronting such a group. The first has its roots in psychology and sociology and is experimental in method. The second has its roots in economics and collective choice and is mathematical and normative in method. The third arises from research on strategic planning in management and the policy sciences and has relied on case studies and the emerging field of systems science for its theoretical underpinnings. In this section, we begin with this last-named research area

*See Nelkin (1977) for a survey of research in the technology area. See also Dodge (1978) for an interesting case history of participation in economic policy formation in Canada. See Ducsik (1981) for an empirically based discussion of citizen participation in electric power system planning.

**For example, Carl Belgie comments in the work by Dodge (1978, p41) that citizen participation is not likely to succeed unless: (i) a crisis is perceived to be imminent; (ii) no subgroup can be blamed for the crisis; (iii) policy makers have plenty of time to devote to the process and have not so committed themselves as to be inflexible to public input; and (iv) the public and policy makers believe that consultation will lead somewhere. Fulfilling all of these conditions in an uncertain and complex environment is clearly a tall order.

since it has the most direct bearing on the issue at hand. The contributions of the other schools of group research are considered in the next section, providing thereafter some concluding comments on open research questions raised by this discussion. Given the widespread interest in groups from various disciplinary perspectives, the issue treated here will necessarily be very selective.

We now turn to the systems theoretic literature on strategic planning and to the work of the two main proponents of this approach, C. West Churchman and Russell Ackoff. First, building on systems theorists such as Ross Ashby, we may note that given the finiteness of any organization's human and technical resources (that is, the finiteness of its potential "variety"), purposeful decision is required as to which aspects of environmental complexity will be monitored and which, perforce, will be left unattended. This gives rise to the definition of strategic planning as the evaluation and choice processes that structure, first, how the world will be viewed (i.e., which of the many possible environmental substructures will be cognitively elevated to recognizable subsystems), and, secondly, the goals that the controlling organization will pursue relative to these subsystems. It is the interaction between this conscious cutting up of the world into subsystems and the specification of goals in terms of these particular subsystems that provide the tension and the substance of strategic planning. Thus the tension between competing interests in strategic planning arises not just because of the goal setting, but also because the subsystems defining the basic background of planning affect what can and cannot be discussed in the planning effort. For this reason, these issues must be solved conjointly. To this end, Ackoff (1974) synthesized a powerful framework for planning which has since been operationalized by Mitroff and Kilmann (1978) using the work of Churchman (1971) on inquiring systems. We now consider the bare outlines of this approach.

Figure 2 illustrates such an approach, following Ackoff (1974). The general idea is that planning is a future oriented activity that requires two essential steps:

(1) specifying an ideal future state (or target) that will serve as a reference point for planning;
(2) determining scenarios that are intended to move the socio-technical system in question from the *status quo* to the desired state.

Performing these two steps requires an explicit analysis of assumptions, viz.,

(3) under what assumptions (on stakeholders, on technology, on the world at large, etc.) are various possible "ideals" actually desirable, and under what conditions will these various ideals be achievable under given scenarios?

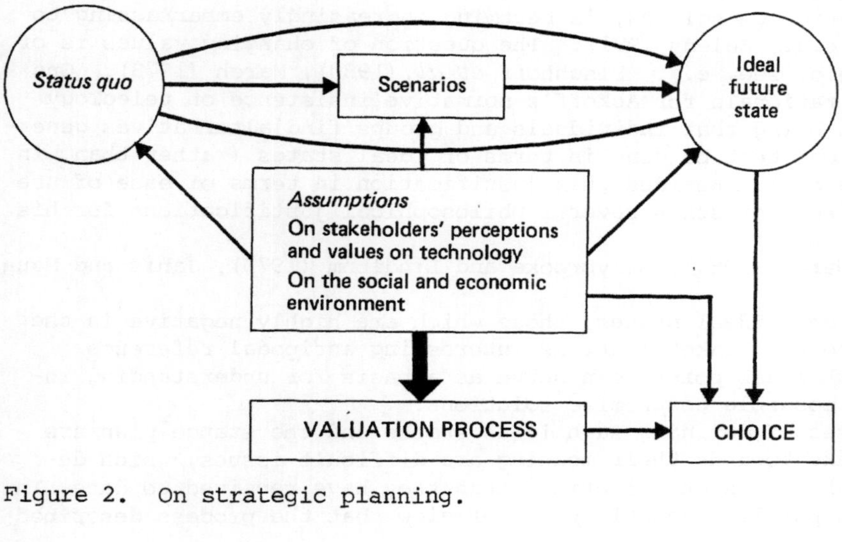

Figure 2. On strategic planning.

Ackoff attaches particular importance to the process by which the above steps are carried out. He argues that most planning methods, especially those rooted in the positivistic analytical tradition, are *reactive* in the sense that they begin with the *status quo* and simply extrapolate this in a locally desirable direction (*reacting*, as it were, to local problems and using only local search). He contrasts this with teleological or purposeful planning, which he refers to as *proactive*, which first determines an "ideal state" in a fairly unconstrained way and then determines scenarios for approaching this ideal state as closely as possible (thus *acting* to attain a *prospective* good). Ackoff's normative assertion is that planning groups are best served by proceeding in the order of steps (1) to (3) above, determining alternatives and feasibility restrictions only after a delineation of possible "ideal states" have clarified what possible directions planning might take. Of course, going through all the bother of designing ideal states may not bring any improvement over, say, standard decision analysis when (as is usually assumed in normative theories of problem solving) all feasible alternatives are known to begin with and stakeholders' preferences over consequences of policy actions are stable. However, when they are not*, the presence of an ideal state can strongly affect search, evaluation, and choice.

Concerning search, there is a large amount of literature indicating that individual and organizational search is local and problematic**, i.e., there is a strong tendency for decision makers to consider only a limited number of alternatives and these are usually dictated as an extrapolation of history and the *status quo* rather than the result of a logical analysis of possibilities. Ackoff's teleological approach to planning is directed explicitly at ameliorating this temporal and technical myopia of planning processes by projecting the perceptions and values of planning participants against an idealized perspective.

Concerning evaluation and choice, research on cognitive dissonance (see, e.g., Festinger 1964) indicates that the presence of an ideal point may strongly influence choice and *ex post* felicity (even when the ideal is not achievable). In commenting on this work, Zeleny (1981) has pointed out*** that ideal states may also be useful reference points for discussion and compromise among conflicting stakeholders.

Ideal points may also serve as a useful reference for understanding the values and decision processes of stakeholders and relating these to external reference groups (as in Thompson 1980). From an ontological viewpoint, the process of defining various ideals as functions of alternative assumptions on stakeholder values is responsive to objections raised by Marcuse (1964) and, in a risk context, by Wynne (1980), that modern technological planning should cool its one-sided ardor for determining the "facts" associated with various policy alternatives and expend some effort in defining the alternative futures against which individual values and technical data become socially meaningful†.

---

*The fact that the generation of alternatives is crucial, and yet rarely studied in normative theories of problem solving, is becoming increasingly embarrassing to decision scientists (see e.g., Zeleny 1981). The question of changing values is of somewhat more recent origin, see, e.g., Fischhoff *et al.* (1980), March (1978). One may view the descriptive rationale for Ackoff's normative insistence on teleology as a basis of planning as being that individuals and groups find alternatives generation and value analysis can best be done in terms of ideal states (rather than, in terms of abstract preferences). Besides this justification in terms of ease of use and understandability, Ackoff presents several philosophical justifications for his proposal as well.

**See, e.g., Cyert and March (1963), Braybrooke and Lindlbom (1970), Janis and Mann (1977).

***Zeleny also discusses anti-ideal states, those which are highly negative in the value system of one or another stakeholder, as interesting antipodal reference points, which together with ideal points can serve as a basis for understanding individual differences and possible compromise solutions.

†Of course, the processes of defining such ideal states and the stance planners and philosophers might take towards their meaning are difficult issues, which deserve a fuller treatment by decision scientists than they have received to date. In this paper, I will take a purely pragmatic point of view that the process described

Although there is little empirical research contrasting Ackoff's method and variants thereof with other methods, there is nonetheless a growing consensus amongst strategic planners and futurologists that scenario construction and idealization offer a viable framework for generating creative alternatives and understanding their assumptional and value referents*. Testing this and other planning methodologies, both in the field and experimentally, appears to be a very fruitful area for future research.

Pursuing the above approach, we can delineate two classes of problems which our planning group may address.

Establishing the Planning Framework

Problem identification and clarification
Design of ideal states
Identification of stakeholders
Scenario generation

Evaluation and Choice

Forecasting: behavioral, factual, and political
Impact and value analysis
Responsibility and accountability analysis

All of the above tasks should be intuitively clear, save possibly the last on responsibility. The issue here is to trace consequences of policy options back through stakeholder impacts and values to determine for various scenarios (e.g., the worst case) who will likely be held responsible for negative outcomes. The purpose of this is not to amplify "anticipatory regret" in Janis and Mann's terminology, but to evaluate the planning process in the light of the incentives it provides through its accountability features for strategic behavior**.

Concerning the above problems, my view is that the systems theoretic tradition has made its most important contributions to the first category of problems: setting the stage for planning. Detailed evaluation and forecasting have been the province of more analytical traditions of decision science, such as theories of collective choice and decision analysis. One may argue that this division of labor is, in fact, appropriate. That is, the broad sweep of human imagination and will should be unleashed in setting the stage for planning, but scenarios and value conflicts generated by this creative process should then be evaluated in a detailed and analytical form to provide realistic feedback on the consequences of alternative plans.

In more detail, the use of groups for forecasting, using Delphi methods or qualitative controlled feedback, is well researched***. Similarly, the methods and problems of dealing with multiple criteria, value trade-offs are reasonably well

---

above may be a useful and understandable way of discussing complex group choice situations with value and perceptual conflicts.

*See, e.g., Mitroff and Kilmann (1978) for an extensive analysis of the various currents that Ackoff and Churchman (1971) have spawned. See also Saaty (1980) for a discussion of a number of case studies using Ackoff's approach within the context of Saaty's hierarchical planning procedure.

**Williamson (1979) provides an interesting discussion of the consequences of responsibility assessment in a regulatory environment (the regulation of food additives). He argues that the interaction between political controls and responsibility assessment are a crucial determinant of agency behavior and suggests, in the light of this, that a formal decision process be agreed upon to allow both intelligent trade-offs to be made as well as to anticipate who will be responsible for legal and stakeholder grievances. In this regard, see also Coppock (1981), and Fincham and Jaspars (1980).

***See Einhorn *et al.* (1977) and Press *et al.* (1979) for a discussion and review of the basic research here. See also the recent survey of the use of such techniques in planning by Hogarth and Madridakis (1981).

understood from an analytic point of view*. However, both forecasting and value analysis require a stage within which these analyses take place. In forecasting, for example, one must begin with a well formulated question. Similarly, for value trade-off analysis among conflicting attributes or stakeholder values, solutions can only be interpreted against the history, the political and social processes, and the assumptions underlying the impact analysis in question. From this perspective, then, the development of scenarios, testing of assumptions, and contrasting of possible ideals is a necessary precondition for applying the more analytic techniques that traditional decision science have provided.

In this regard it is important to note the important contributions of Richard Mason and Ian Mitroff**, who proposed a synthesis of Ackoff's work on teleological planning with the pathbreaking work of Churchman (1971) on inquiring systems. The nature of their proposal was to abstract the methods of argument and evidence generation from various philosophical traditions and use these contrasting methods for various elements of the above tasks of generating ideal states, and evaluating scenarios and assumptions. For example, following Hegel, one can imagine a "dialectical method" of assumptional analysis. By generating two antithetical sets of assumptions (e.g., assuming that the development of some protective device will be fast or slow), one can generate contrasting scenarios, which can be used to elucidate both the scope of feasible scenarios, and the implications of the antithetical assumptions at issue. In a similar manner, an analytical (Leibnitzian) approach would properly be used to explore the implications of an interrelated set of technical assumptions in a field whose technical scope was well understood. By using various approaches to generate and measure the scope of assumptions and associated scenarios against ideal states, the Mason-Mitroff framework provides an interesting operational extension of the Ackoff-Churchman approach.

As a final point on the systems theoretic tradition, the recent work of Saaty (1980) is important in that it integrates the Ackoff ideal-scenario-assumption perspective with more analytical (multicriteria) methods of scenario evaluation and choice. Saaty argues that there is usually a wide gap between traditional analytic methods of evaluation and the problem representations which human planners find natural (i.e., easily comprehensible). He suggests that to narrow this gap, one should use a problem representation which relates easily to the underlying scenarios and assumptions in terms of what human planners think. For this purpose, Saaty proposes what is really a variant of decision trees, with the consequences of various decisions represented in multicriteria fashion. He then proposes an operational method, with nice analytical and empirical properties, that uses the hierarchical (or, if you prefer, tree-like) scenario and value structures as a basis for eliciting the preferences of individuals and groups for desired scenarios in terms of their multicriteria consequences. A general critique of the appropriateness of Saaty's methods and its comparative strengths and weaknesses awaits future research. Its basic point of linking the underlying framework (of scenarios and ideal states) of planning to analytical processes for evaluating specific options is clearly an important step in the right direction.

4. OTHER RELEVANT GROUP RESEARCH

We now turn to traditional group research and to analytical methods in the collective choice area as these relate to the problem of group planning. Concerning both of these, the discussion will be quite brief, pointing only to their implications for planning in the risk area.

The roots of experimental group research go back to Lewin's research on group dynamics in the late 1940s***. The primary concern of Lewin and his co-workers was

*See, e.g., Lasswell (1971), Keeney and Raiffa (1976), Saaty (1980) and Zeleny (1981) for a representative sampling of the literature on multicriteria problems in theory and practice.

**See Mason and Mitroff (1973) and the subsequent developments in this area described in Mitroff and Kilmann (1978).

***See Shaw (1971) and Cooper (1975) for overviews of Lewin's original research and the issues it spawned.

the process by which group structures were formed and how these contributed to learning and to individuals' relations to groups. These ideas were taken up immediately by psychiatrists and clinical therapists as basis of group therapy. Out of this and the sociological tradition of referent groups grew several inter-related strands of group research, which we now briefly consider*.

4.1. Phenomenology

The concern here has been to identify existential characteristics of groups (cohesion, trust) and to relate these to subjective perceptions of group members (belongingness, communion). The reason phenomenological theories are important to planning is in shedding light on the relationship between group level variables and member perceptions and motives**.

4.2. Group Structure

This research has attempted to clarify the relationship between group task, communications structure (who can talk to whom and how easily), group size and composition (e.g., male, female, or mixed), and group performance. MacKenzie (1976) and Steiner (1972) provide extensive discussions of the research in this area. MacKenzie verifies experimentally that something like a "fit" equation obtains amongst these variables in the sense that if one is changed (e.g., task structure) by the experimenter, then other aspects of group structure and environment under the control of the group (e.g., the communications structure) will change in predictable ways. It appears that there is something like a natural or ideal structural condition, dictated by the needs of the group and the task structure confronting them, towards which group structure tends when left free to vary. The point of this from a planning perspective is that information processing and the communication resources of a group should be compatible with this natural fit point to avoid internal structural pressures.

4.3. Group Process and Learning

This area has been concerned with decision processes, norms, and communication patterns within groups as a function of group size and structure, and procedural interventions. The general literature here is covered by Cathcart and Samovar (1979) and by Argyris and Schon (1974). We may envisage the general point of these studies as attempting to understand how variables, such as communication, trust, and cooperation, interact. One may view the major prediction of this research as being that (ease of and level of) communication, trust, and cooperation are mutually reinforcing. For example, Marwell and Schmitt (1975) and Dawes (1980) provide interesting predictions of when cooperation and trust may be expected. This research corroborates the idea that certain objective characteristics of the problem situation (e.g., the ability of an individual to profit at the expense of the group) in their interaction with the ability of the group to exercise sanctions for "deviant" behavior act as fundamental determinants of trust and cooperation. Communication tends to reinforce both cooperation and the ability to design creative compromise solutions (see Steinfatt and Miller 1974, Dawes 1980). This literature has also emphasized the consequence of lack of trust on individual processes as these interact with group processes. For example, the research of Argyris and Schon (1974) points out that the natural consequence of lack of trust is defensive behavior, relying on defensible and accepted theories, with a consequent lack of openness to creative problem solving. A further consequence of lack of trust is unwillingness to accept the

---

*See the editorial introduction to Cooper (1975) for a brief history of the roots of modern group research. We follow here the taxonomy given by Shadish (1981).
**See Shaw (1971) for a research summary of this area.

responsibility for the actions of the group even when one's own (explicitly espoused) policy alternatives are adopted*. This research clearly indicates the fragile nature of trust, communication, and creative problem solving. One direct implication for group problem solving, which Zeleny (1981) suggests but which has not yet been tested experimentally to my knowledge, is that trust and cooperation are promoted in a planning environment by providing appropriate decision-aiding technology (easing communication, as it were) to allow easy exploration of compromise solutions.

4.4. Leadership Research

Finally, a good deal of group dynamics and social psychological research has been devoted to studying the role of the group leader on group performance and satisfaction experienced by group members. Knowledgeable and caring leaders who exercise a strong, but not manipulative role, have had a significant impact on group performance in experimental and clinical settings (see Steiner 1972).

Now let us turn our attention briefly to the analytical tradition of group choice that has been studied under the heading of collective or social choice (see, e.g., Sen 1970, Pattanaik 1978). The collective choice literature generally starts with a set of agents, a set of alternatives, and, for each agent, a well defined preference order over the alternatives. The object of this theory is to determine some general procedure for determining a "socially superior" subset of the given initial set of alternatives. For example, one such procedure would be to order the alternatives in some fashion and then to use a majority rule voting procedure as a pairwise elimination procedure, until only one alternative remained. Unfortunately as Arrow showed in his famous paradox, the alternative selected by this procedure will depend on the order (or agenda) in which alternatives are compared. Moreover, if one restricts one's attention to procedures which satisfy certain reasonable properties, this defect is not remediable. Indeed, there is a disappointingly large and growing list of negative results on the possibility of constructing social choice mechanisms satisfying reasonable criteria, at least if we take preferences and alternatives as given and seek a semi-automatic procedure for accomplishing choice.

The above noted problems only get worse when we introduce elements of uncertainty as in Bacharach (1975). Under these conditions, it takes very strong assumptions to ensure even that a reasonable consensus on subjective beliefs can be obtained, let alone the definition of a group choice function based on such beliefs. Essentially, to obtain the existence of such a group choice function, one must *assume* that it exists (in which case it turns out to be nicely representable as a linear sum of the utility function representations of individual preferences**). Now, such a procedure is not necessarily inappropriate, that is, if one is looking for a collective decision, assuming that one can in fact reach consensus may be a constructive beginning. The problem, at least for the risk area, is that such consensus is not likely to be promoted by taking alternative preferences and beliefs as given and mechanically searching for an optimum within these given details using a fixed procedure. Even if we had such a notion in mind, the literature on collective choice is quite clear that the search for a reasonable procedure would be in vain. Rather, consensus would seem better served through planning and participative group process to allow conflicting beliefs and value structures to be woven into constructive compromise.

---

*For a review of recent research responsibility and attribution theory, see Fincham and Jaspars (1980).

**See Dyer and Sarin (1979) for a recent discussion and extension of Harsanyi's original results in this regard.

5. CONCLUSIONS

The questions raised by the above discussion seem to fall in three areas.

(1) Linking policy planning groups and stakeholders to the broader social context of their planning (the political design question). We have had little new to say about this. However accomplished, this step has to be understood as a major link between planning groups and "reality".

(2) Improving the performance of policy planning groups in accomplishing specific planning tasks. Research on group problem solving suggests that such matters as group structure, communication and decision aids, and the ability to generate easily compromise solutions, all contribute to this process. It has also been argued that the systems theoretic approach provides an often missing ingredient to this normative process by helping to delineate the assumptional and value background against which policy analysis is done.

(3) Finally, the major challenge for future research seems to be in linking the unstructured processes of ideal state design and scenario generation with the analytical evaluation of such scenarios. While we wish to unleash the powers of group imagination, we also wish to restrain our flights of fancy by the real constraints which technology and social processes impose. An argument can be made that this linking of unstructured design with structured evaluation can be significantly improved using computer based decision support systems*. In any case, this linking process is, in my opinion, the key to effective group planning and policy analysis.

REFERENCES

Ackoff, R.L. (1974) Redesigning the Future, A Systems Approach to Societal Problems, New York: Wiley.

ACIR (Advisory Commission on Intergovernmental Relations)(1979) Citizen Participation in the American Federal System. Report A-73. Washington, DC: US Government Printing Office.

Alter, S. (1980) Decision Support Systems: Current Practice and Continuing Challenges. Reading, Mass.: Addison-Wesley.

Argyris, C., and Schon, S. (1974) Theory in Practice. San Francisco: Jossey-Bass.

Arrow, K.J. (1970) Political and economic evaluation of social externalities, in The Analysis of Public Output, edited by J. Margolis. Washington, DC: National Bureau of Economic Research.

Bacharach, M. (1975) Group decisions in the face of differences of opinion. Management Science 22:182-91.

Bauer, V., and Wegener, M. (1977) A community information feedback system with multiattribute utilities, in Conflicting Objective in Decisions, edited by D. Bell, R. Keeney, and H. Raiffa. New York: Wiley-Interscience.

Belardo, S., and Wallace, W.A. (1981) The Design and Test of a Microcomputer Based Decision Support System for Disaster Management. Working Paper. Troy, N.Y.: Rensselaer Polytechnic Institute, School of Management.

Berkowitz, L., and Walster, E. (1976) Equity Theory: Toward a General Theory of Social Interaction, Vol. 9 in the series, Advances in Experimental Social Psychology. New York: Academic Press.

Braybrooke, D., and Lindblom, C. (1970) A Strategy of Decision. New York: Free Press.

Cathcart, R.S., and Samovar, L.A. (1979) Small Group Communication. Dubuque, Iowa: Wm. Brown.

---

*See Alter (1980) for general background on decision support systems and Belardo and Wallace (1981) for a specific case study of the use of DSS in a risk management context.

Conrad, J. (1980) Society, Technology and Risk Assessment. New York: Academic Press.
Cooper, C.L. (Ed.)(1975) Theories of Group Processes. New York: Wiley.
Coppock, R. (1981) A Conceptual Framework for Managing Technological Hazards. Berlin: International Institute for Environment and Society, Science Center.
Cyert, R.M., and March, J.G. (1963) A Behavioral Theory of the Firm. Englewood Cliffs, NJ: Prentice-Hall.
Dawes, R.M. (1980) Social dilemmas. Annual Review of Psychology 31:169-93.
Dodge, W. (Ed.)(1978) Consultation and Consensus: A New Era in Policy Formation. The Conference Board in Canada, December.
Ducsik, D.W. (1981) Citizen participation in power plant siting: Aladdin's lamp or Pandora's box? Journal of the American Planning Association, April:154-166.
Dyer, J.S., and Sarin, R.K. (1974) Group preference aggregation rules on strengh of preference. Management Science 25(9).
Einhorn, H.J., Klemper, E., and Hogarth, R.M. (1977) Quality of group judgement. Psychological Bulletin 84(1):158-72.
Eiser, J.R., and Van der Pligt, J. (1979) Beliefs and values in the nuclear debate. Journal of Applied Social Psychology 6:524-36.
Festinger, L. (1964) Conflict, Decision, and Dissonance. London: Tavistock Publications.
Fiddle, S. (Ed.)(1980) Uncertainty: Behavioral and Social Dimensions. Praeger.
Fincham, F.D., and Jaspars, J.M. (1980) Attribution of responsibility: From man the scientist to man as lawyer, in Advances in Experimental Social Psychology, Vol. 13, edited by L. Berkowitz. New York: Academic Press.
Fischhoff, B., Lichtenstein, S., Slovic, P., Derby, S., and Keeney, R. (1980) Approaches to Acceptable Risk: A Critical Guide. Report to the Nuclear Regulatory Commission.
Fischhoff, B., Slovic, P., and S. Lichtenstein (1980) Labile values: A challenge for risk assessment, in Society, Technology and Risk Assessment, edited by J. Conrad (1980).
Greer-Wooten, B. (1980) Context, concept and consequence in risk assessment research: A comparative overview of North American and European approaches in the social sciences, in Conrad (1980).
Hogarth, R.M., and Madridakis, S. (1981) Forecasting and planning: An evaluation. Management Science 27(2):115-38.
Janis, I., and Mann, L. (1977) Decision Making. New York: Free Press.
Keeney, R.L., and Raiffa, H. (1976) Decisions with Multiple Objectives. New York: Wiley.
Kleindorfer, P.R., and Kunreuther, H. (1981) Descriptive and prescriptive aspects of health and safety regulation, in The Benefits of Health and Safety Regulation, edited by A.R. Ferguson and E.P. LeVeen. Ballinger.
Kunreuther, H., Ginsberg, R., Miller, L., Sagi, P., Slovic, P., Borkan, B., and Katz, N. (1978) Disaster Protection: Public Policy Lessons. New York: Wiley Interscience.
Lasswell, H.D. (1971) A Pre-View of Policy Sciences. New York: Elsevier.
Lave, L. (1972) Risk, safety and the role of government, in Perspective on Benefit-Risk Decision Making. Washington, DC: National Academy of Engineering.
Lawless, E.W. (1977) Technology and Social Shock. New Brunswick, New Jersey: Rutgers University Press.
Luft, H.S. (1976) Benefit cost analysis and public policy implementation: From normative to positive analysis. Public Policy 24(4):437-62.
MacKenzie, K.D. (1976) A Theory of Group Structure. New York: Gordon and Breach.
March, J.G. (1978) Ambiguity and the engineering of choice. The Bell Journal 9(2).
Marcuse, H. (1964) One-Dimensional Man. Boston: Beacon Press.
Marwell, G., and Schmitt, D.R. (1975) Cooperation: An Experimental Analysis. New York: Academic Press.
Mason, R.O., and Mitroff, I.I. (1973) A Program for Research on Management Information Systems. Management Science 19(5):475-87.

Mitroff, I.I., Barabba, V.P., and Kilmann, H. (1977) The application of behavioral and philosophical technologies to strategic planning. Management Science 24 (1):44-58.

Mitroff, I.I., and Kilmann, R.H. (1978) Methodological Approaches to Social Science. San Francisco, California: Jossey-Bass.

Nelkin, D. (1977) Technology and public policy. Chapter 11 in Science, Technology, and Society, edited by I. Spiegel-Roesing and D. de Solla Price. Sage Publications.

O'Hare, M. (1977) Not on my block you don't: Facility siting and the strategic importance of compensation. Public Policy 25:407-58.

Otway, H.J. (1980) Perception and acceptance and environmental risk. Zeitschrift für Unweltpolitik 2:589, 593-616.

Pattanaik, P.K. (1978) Strategy and Group Choice. Amsterdam: North-Holland.

Press, J.S., Ali, M.W., and Yang, C-F.E. (1979) An empirical study of a new method for forming group judgements: Qualitative controlled feedback. Technological Forecasting and Social Change 15:171-89.

Ronge, V. (1980) Theoretical Concepts of Political Decision-Making Processes, in Conrad (1980).

Saaty, T.L. (1980) The Analytic Hierarchy Process. New York: McGraw-Hill.

Sen, A.K. (1970) Collective Choice and Social Welfare. San Francisco, California: Holden-Day.

Shadish, W.R. Jr. (1981) Theoretical observations on applied behavioral science. Journal of Applied Behavioral Science 17(1):98-112.

Shaw, M.E. (1971) Group Dynamics: The Psychology of Small Group Behavior. New York: McGraw Hill.

Starr, C., and Whipple, C. (1980) Risks of risk decisions. Science 208(6).

Steiner, I.D. (1972) Group Process and Productivity. New York: Academic Press.

Steinfatt, T.M., and Miller, G.R. (1974) Communication in game theoretic models of conflict, in Perspectives on Communication in Social Conflicts, edited by G.R. Miller and H.W. Simons. Englewood Cliffs, NJ: Prentice-Hall.

Thompson, M. (1980) An Outline of the Cultural Theory of Risk. WP-80-177. Laxenburg, Austria: International Institute for Applied Systems Analysis.

Williamson, O.E. (1979) The anguish of saccharin: The decision process approach and its alternatives. Discussion Paper No. 53. Pennsylvania: University of Pennsylvania, Center for the Study of Organizational Innovation.

Wynne, B. (1980) Technology, risk and participation: On the social treatment of uncertainty, in Conrad (1980).

Zeleny, M. (1981) Descriptive decision making and its applications. Applications of Management Science 1:327-388. JAI Press Inc.

Zerbe, R., and Urban, N. (1980) Toward a Public Interest Theory of Regulation. Conference Paper No. 70. National Bureau of Economic Research, Inc.

Market Risk Assessment of Catastrophic Risks

William B. Fairley
Analysis and Inference Inc., Boston, Massachusetts, USA

1. INTRODUCTION

This paper is intended to encourage interest in two different and separable conjectures about the roles that properly established markets might play in improved social management of catastrophic risk. First, the accuracy, credibility, and usefulness of risk assessments for catastrophes could be improved if they were made the tools of risk producers with a direct financial stake in their accuracy.

Second, although not the main theme of this paper, a variety of advantages—some economic, some broadly political, and some perhaps philosophical—accrue in a risk system in which risk producers are accountable to the fullest practical extent financially for the risk liabilities they impose on others. Financial accountability can be required with more or less governmental involvement in the mechanics of compensating victims and for property losses. A variety of proposals have come before the US Congress and state legislatures that would mandate financial responsibility requirements for corporations (e.g., HR 1414 (1980), Massachusetts S 742 (1980) for liquefied natural gas (LNG) and liquefied petroleum gases (LPG) hazards HR 6390 (1980), HR 789 (1980) for nuclear power hazards). Some types of public involvement appear to help to establish a well functioning market and have a coherent rationale under a market approach. Others, however, appear to constitute an incoherent intervention into a market mechanism without substitution of adequate public institutions for control. Considerably more analysis of these kinds of proposals is required before conclusions about their likely effects can be made.

Financial responsibility requirements can take a variety of forms, including self-insurance, commercial insurance, industry risk-pooling associations, and government-run compensation funds with subrogation rights to risk producers. The choice of the form of finance can be vital to the performance of a risk system along all dimensions of public interest including the degree to which control of risk is achieved, the acceptability of risk, the efficiency with which safety is produced (or risk avoided), the extent to which individuals' preferences about risk are taken into account in decision making, and administrative feasibility and costs. In particular, the role of private insurance markets *vis à vis* that of government requires scrutiny in the light of current literature (for example, see Kunreuther and Pauly 1981, and references therein).

The point of view taken here is not to recommend specific roles for markets in controlling catastrophic risks and in providing accurate assessments of such risks, but rather to suggest possibilities. Decision making in a market context has at least superficially attractive possibilities. A deeper study of the potential role of markets would rest on a general theme developed by others (see, for example, Schultze 1977) that even a "market failure" does not in and of itself create a presumption of favoring government enterprise or detailed regulation, since there is certain knowledge that the latter will be better. Market frameworks, however, which are appropriately maintained by government, with adequate responsibility clearly and adequately assigned to risk producers,may turn out to perform well.

2. CATASTROPHIC RISKS AS COSTS OF PRODUCTION

Accidents occurring in the course of industrial activity are considered by economists to be costs of production. Under standard assumptions of economic theory, and under widely shared presumptions, good things happen in the economic world when costs of production are borne by the producers. In particular, if they bear the costs, producers have financial incentives to reduce them. Further, if producers bear the costs initially, then final consumers who purchase and enjoy the use of the product bear this burden (since it is included in the price), rather than persons who did not choose to purchase and enjoy it.

Most accidents are counted routinely in the costs of production—things like normal breakage, unfortunate occupational accidents, etc. Accidents on the scale of chemical plant explosions, while exceptional, are viewed as part of the general cost of being in business. Commercial insurance may cover the financial costs of such events. In all of these cases producers can generally be assumed to have the usual financial incentives to cut costs—either of the accidents themselves or the insurance premiums that cover the costs incurred in the event of an accident. In cutting costs producers of course cut accidents. Further, producers are generally presumed to know best how to cut costs. They may choose more careful organization under existing technology or sponsor innovations that will enhance safety and reduce risk.

Parenthetically, let me add that in relating these standard presumptions I am deliberately sliding over numerous caveats and potential research areas for economic specialists such as questions of informational perfection in the market for insurance and questions relating to perfect markets for innovations (for these discussions see the survey by Hirshleifer and Riley 1979). Rather, I rely only on whatever support is obtained under simple certainty models plus the general assumptions that are employed in industrial organization economics and in business policy about the good results of markets in which costs of production are borne by producers.

When we come to catastrophic accidents, however, we find that a very different situation exists in business practice and in public policy towards business. Catastrophic accidents on the scale of a huge fire generated in downtown Boston from an LNG tanker accident, or a meltdown in a nuclear power plant, are not currently treated in the same way as most smaller accidents. The main difference is that catastrophic risk is a cost of production that is typically borne only in part—and sometimes this is a small part—by producers. Risk (variously described as a spillover or an externality of production) is typically largely borne by persons or entities other than the risk producer. Exposed persons in the geographic vicinity absorb part of the costs of production and provide what might be called a risk subsidy towards the production in question.

The full costs of catastrophic risks are not generally borne by producers because their assets and insurance coverage are not sufficient to cover the potential costs of credible accidents. Even large corporations might be bankrupted by these liabilities. However, the ownership structure for potential catastrophic activities (as detailed, for example, for LNG in the US General Accounting Office (1978) study of LNG risks) shields the larger parent corporations from financial responsibility for catastrophic liabilities through the use of subsidiaries. In the case of nuclear power, the US congress has expressly limited total public and private liability for a nuclear accident to $560 million through the Price-Anderson Act of 1957 (amended twice since 1957 with ten-year extensions, but with no change to the upper limit of recoverable damages).

Regulatory authorities (e.g., the Federal Energy Regulatory Commission for LNG and the Nuclear Regulatory Commission for nuclear power) review site applications and grant licenses for the construction of new facilities with a catastrophic risk potential. Through this avenue producers may be enabled to build facilities at lower cost than they otherwise would—with risk subsidies provided by persons exposed to the risks.

On the other hand, since the present system does not guarantee that anyone has a responsibility for making a comparison between what are probably the most important alternatives with respect to risk costs—namely site regulation—is likely to be inefficient in its choice of safety technology. An example would be not trading off more remote site locations against hardware fixes at sites near population centers.

Producers will choose sites that public authorities permit and that minimize their costs. Since these costs include only a part of the catastrophic risk costs they may not minimize these risks. They will have no direct incentive to search for the least-cost site solution that takes catastrophic risks into account.

The mixed public-private nature of the present system, in failing to assign clear responsibility for financial liabilities and for risk management, may exemplify the concern of Nelson (1981) about a system of economic organization that combines the worst features of public and private systems for responsibility. Failure to assign responsibilities may be viewed as a failure of public authority to establish properly the framework required for private markets to operate well. It is perhaps difficult to say whether what is involved is "market failure" or "nonmarket failure" to use the phrase of Wolf (1979). The issue is therefore not simply between more or less regulation; rather it is between a good public foundation for well functioning private systems and a poor public foundation. Nor is the issue between public and private interests. Private interests stand to gain something in either case, but it would be preferable if certain private interests, such as persons living near hazardous facilities, were not sacrified.

In summary, all the legal, statutory, and regulatory frameworks for assigning responsibility for the costs of catastrophic risks in the USA today permit or even endorse the concept that those costs need not be borne by the risk producer and can properly be absorbed by the persons exposed to those risks.

The current system violates the norm of standard economic principle and practice in the assignment of costs. It follows that we should not expect the advantages such as least-cost incentives for risk control, and the equitable allocation of costs onto persons who purchase and enjoy products, that are generally believed to follow from adherence to that norm. But our main theme is assessments of catastrophic risks. How does the current system affect these?

3. MARKET RISK ASSESSMENT: CREDIBILITY, USEFULNESS, AND ACCURACY

First, the credibility of risk assessments is a casualty of the process by which some parties stand to gain (and others stand to lose) very substantial sums depending upon the influence they can exert on regulatory bodies. Risk assessments are usually sponsored by parties with financial interests clearly identified with findings of low or high risk. Applicants before regulatory authorities who seek construction licenses for hazardous facilities clearly have a financial interest in low-risk determinations that would speed licensing approval, while resident groups near proposed sites just as clearly have financial interests in high-risk determinations that would send the facility off to another location. Under the circumstances in the present system, participants in regulatory hearings routinely discount the accuracy and credibility of risk assessment done for the other parties. Besides spawning a substantial amount of cynicism about the reliability of political processes, the outcome leaves responsible decision makers without as sound a basis for making a decision as might be possible.

We can imagine improving the situation with an injection of disinterested objective risk assessment. However, while possibly better than nothing, the availability of objective, accurate and credible assessment would not in any way change the underlying problem of misassignment of effective financial responsibility of risk, reduced incentives for risk control, and an inequitable burden of risk. Risk assessments introduced after a site has been selected are of extremely limited usefulness because a decisive variable influencing risk—the location—has been previously determined (see Mandl and Lathrop 1981). There are some tremendous incentives to find estimated probabilities within a range of acceptable risk of alternatively acceptable risk within the range of probabilities estimated!

If he is financially responsible for the risk, the risk producer will have a greater interest in accurately estimating the probabilities and costs of accidents and in least-cost choices of methods to reduce them. For example, if an expected liability from an accident or the insurance premium to cover the liability could be reduced sufficiently by moving the site of a hazardous facility to a less densely populated location, a direct financial interest will come into play to aid the adoption of that less risky alternative. The costs of liability to the risk producer can be viewed as part of the demand for safety on the part of potential hazard victims. Accurate demand estimation, like accurate cost estimation for risk reducing technology or design, is in the risk producer's interest.

Risk assessments by risk producers that are too optimistic lead to losses or expected losses to themselves. Assessments that are too pessimistic lead to profitable opportunities foregone or, in workably competitive markets, to noncompetitive cost estimates. In either case the risk producer stands to lose—but only if the costs of risks are really made part of his production costs. Risk assessments made by assessors in workably competitive markets who are responsible for risk costs can be called "market risk assessments". Further, it is useful to refer to a "responsibility-based" risk system as one in which risk producers are financially responsible for risk costs and therefore make market risk assessments.

4. MARKET RISK ASSESSMENT: OTHER FEATURES

Consideration of market risk assessments throws light on some other issues. Consider, for example, the "nebulosity problem". Probabilities of rare events often appear nebulous. The conditions under which the events could occur are difficult to specify with any precision or completeness so that a probability is difficult to envisage even conceptually (see Fairley 1981). Estimates of probabilities seem nebulous for the same reason and also because neither a record of experience nor a good model is available. Such nebulosity creates difficult problems for public decision makers because their decisions, which are written and issued under an official imprimatur, will inevitably seem to be built on sand. There is a great demand for certainty on the part of the public—a demand that cannot honestly be met under the circumstances. Responsibility-based risk systems would appear to deal with numerous other business risks, a large number of which are nebulous, such as market demand shifts, and the productivity of research and development programs. With lower visibility the hunches and intuitive reasoning that can be every bit as important as more formally gathered information and analysis in market assessments are not subject to unfair public comparison against unrealistic standards for accuracy. The risk of financial loss for mistaken assessments provides a direct financial stake in achieving achievable levels of accuracy at a reasonable cost, regardless of the nebulosity of the event being estimated.

Consider the "politics versus rationality" problem. Risk assessments based on "objective" outcome measures such as number of fatalities, or on the rational choice paradigm for decision making, are often contrasted with assessments that take into account a variety of psychological or subjective factors such as—in Lathrop (1981), for example—dread of catastrophes or the worst possible consequence. Political decision making is seen as responsive to psychological or subjective factors in a way that is not handled within rational prescriptive models for decision making.

Responsibility-based systems appear to be less based on a dichotomy between an individual's "rational" and, by implication, what are often viewed as "irrational" sources of preference about risk: insofar as individual's preferences about risk play a role in the compensation they demand or receive for risk liabilities, then these preferences will influence decision making about risks in these systems regardless of whether others view them as rational, irrational, or something else.

Consider the "commensurability" problem. Actual decision making, about hazardous activities involves the consideration of a large number of factors besides threats to life, limb, and property, and yet risk assessments under the present system deal almost exclusively with risks to life alone, excluding injuries, property loss, etc. Risk assessment as practised to date has trouble incorporating all these seemingly incommensurable considerations in a single decision making framework, and yet the interdependencies of all these various decisions are widely recognized.

In a responsibility-based system a risk producer confronts the financial cost of compensation for risk liabilities and incorporates this money cost into all the other considerations that go into the business decision. In this customary process of business decision making, a variety of ends are valued commensurately. Aiding the customary process by supplying, as one input, new and better risk assessments—whether limited to a focus on fatalities or not—can generally be expected to improve it. However, using risk assessments limited to a consideration of fatalities in a system that lacks financial incentives to take other factors into account appears to be less likely to improve the system overall.

5. CONCLUSION

The purpose of this paper has been to advance some speculative reasons why risk assessments made by risk producers who are financially responsible for costs of catastrophic risk might be more credible, useful, and accurate and enjoy certain other advantages over present risk assessments presented in regulatory hearings and other public decision contexts. No firm conclusions are possible, if only because important issues of institutional design and of practical functioning of the markets discussed have not been addressed. However, the advantages in theory of what are called herein "market risk assessments" in "responsibility-based" systems of risk assignment and control seem attractive enough to warrant further exploration.

REFERENCES

Fairley, W.B. (1981) Assessment for catastrophic risks. Risk Analysis 1(3): to appear.

Hirshleifer, J., and Riley, J.G. (1979) The analytics of uncertainty and information—an expository survey. Journal of Economic Literature December:1375-421.

Kunreuther, H., and Pauly, M. (1981) Equilibrium in Insurance Markets with Experience Rating. IIASA Preliminary Paper, March 1981. Laxenburg, Austria: International Institute for Applied Systems Analysis.

Lathrop, J.W. (1981) Evaluating Technological Risks: Prescriptive and Descriptive Perspectives. This volume.

Mandl, C.J., and Lathrop, J. (1981) Assessment and Comparison of Liquefied Energy Gas Terminal Risks. WP-81-98. Laxenburg, Austria: International Institute for Applied Systems Analysis.

Nelson, R.R. (1981) Assessing private enterprise: an exegesis of tangled doctrine. Bell Journal of Economics, Spring issue:93-111.

Schultze, C.L. (1977) The Public Use of Private Interest. New York: Brookings Institution.

US General Accounting Office (1978) Liquefied Energy Gas Study, Vol. 1, July.

Wolf, C. (1979) A theory of nonmarket failure. Journal of Law and Economics, April.

Chapter 5

GENERAL DISCUSSIONS AND FUTURE RESEARCH NEEDS

General Discussions

The afternoon sessions of the Summer Study were devoted to discussions of a more general nature. The participants were divided into four groups and were given a set of questions for discussion that had been presented on the first day as part of the general IIASA overview.

The material that follows comprises transcripts of two plenary session discussions that took place after the small group meetings. We have listed some of the questions which we feel stimulated interaction among participants. We hope that the ideas which emerged from these open interactions will set a tone for future research on institutional and decision process aspects of risk.

Session I

- How can analysis, given its limits, be more useful in generating and evaluating alternatives?
- How are analysts currently involved in the policy process?
- How can analysts play a more constructive role in interactive ongoing decision processes?
- In the absence of formal risk assessment how have societies in the past handled risk successfully?
- Why is risk treated as a separate entity in the decision process?

Schoemaker: One thing I noticed is an interest in the sociology of risk assessment; specifically, where do experts who propose to help society to make decisions come from? It seems that we do not fully understand, at least from a cross-cultural viewpoint, how those sociologies differ. Initially, there seemed to have been a vacuum and suddenly experts come into existence because a technology imposes itself on us. It seems that various people in the discussion groups are interested in what one may call the sociology of the risk assessment experts, i.e., its subculture.

Kasperson: A quite striking issue in this general discussion is how to get a legitimate authoritative credible risk assessment, i.e., one that will be accepted as that by a variety of different parties. Obviously there is concern that the present process does not achieve that.

Douglas: In our group we tried to quiz the German representatives as to how people cope without this kind of analysis. There must have been some vacuum in certain countries but not in others, which caused this type of analysis to come forth where it did. One question that we might look at, apart from ones already on board, is the macro-national structures in which risk analysis as distinct from decision analysis has been utilized.

Kunreuther: You are saying that if you look at these different cultures you perceive that some of them may have felt that risk analysis was an appropriate tool to use, while others avoided it, until later on in the process.

Douglas: That is interesting because the IIASA representative in our group, Joanne Linnerooth, was saying that of course everyone needs to work out how to make these risk decisions, but we also heard that people did not always have to do that.

Kunreuther: We heard a story connected with that at our LEG workshop. One of the people from the California Coastal Commission actually raised the question himself:

"Should I ever hire a consultant to do a decision analysis for me?" He himself had been trained in this technique and said they made some calculations on the back of an envelope but kept the figures in a drawer in case they had to justify their actions. He would never want to make the analysis public or have a consultant undertake them because they would be looked at in an entirely different way. So some of this may actually take place even though it is not made explicit. Suppose we looked at a country where risk analysis was not used. What are the characteristics of the cultural and the historical background which created an atmosphere where no explicit quantitative assessment was undertaken?

Douglas: English history seems to have been one of ups and downs in this respect, with an initial resistance. In 1972 the Roskill Commission caused public and official confusion by inserting fictional costs into the calculations in the report on the decision of where to locate a third London airport. Volumes have been written to say that after this no one can possibly use cost-benefit analysis except in very narrow technical engineering situations. Yet ten years have gone by and Joanne Linnerooth tells me it is still in good repair so that people are still using it. There must be some political needs which it meets which are absent in West Germany, where it is not used.

Schütz: We do not use it very much and it should be noted that there can be no explicit trade-off between human lives and economics. I think it is quite accepted that normally no one would want that kind of trade-off because it is unethical.

Kleindorfer: What about cost effectiveness, i.e., considering not the trade-off between dollars and lives, but rather the dollar cost of the last life saved—perhaps you cannot do that either?

Schütz: You can do it if, for example, you consider nuclear power plants where you will have some technical differences between the power plants. You could investigate, for example, whether a different emergency cooling system would be more or less effective for the same amount of money, but you will not explicitly calculate lives saved or not saved.

Wynne: In answer to Howard Kunreuther's question, and perhaps you could correct me if I am wrong, I have the impression that nobody is conducting any formal risk assessment of information technology.

Kasperson: The Swedes have done a big risk assessment on computers and privacy.

Stoto: There have been studies and similar investigations in the United States, but not formal in the sense of setting out quantitative risk.

Wynne: I would not equate formal with quantitative, but I would suggest that it is for the same reason that it was not done for nuclear power in the 1950s.

Lathrop: I think this is a very interesting line of discussion. I would suggest that there are two reasons why we have not yet seen formal risk assessment for information technology. One is that risk assessment is part of a process especially sensitive to the image or ideation of hazard. Hazard in the nuclear industry or the space program involves a very visible or identifiable menance, i.e., you cannot describe in one or two words or images the hazards of information technology. You might say one of them is privacy—or loss of privacy—but in fact the set of hazards is much more complicated. Lacking an easily visualized hazard, there is no disposition toward risk assessment with information technology. The second reason is related to the first: there is no easily characterized single dimension that captures very much of people's concerns about information technology. Nuclear safety is amenable to risk assessment because there is a natural format for risk. People naturally see that risk in terms of lives lost.

Drake: I think if you look at the roots of risk analysis it started with reliability analyses for the space program and then became a tool in the design of control systems for complicated facilities,i.e., chemical plants, nuclear plants. It evolved as a design tool rather than a tool for analyzing potential for major catastrophes. For systems where somebody had to sit down and design them and feel comfortable that they would work adequately, risk analysis provided a methodology for systematic evaluation of safety and operability. When risk analysis started to be used for looking at the rare event, it changed character. Then it was no longer a matter of just putting in statistical data, but also of fitting subjective judgments, biases, etc., into the same matrix. Like a computer, if you put garbage in, you will get garbage out. But, if you come out with a stack of computer output, people tend to believe somehow that this is blessed because it looks official and scientific. Certainly it can be misused. I think when you get into the softer areas there is an important question as to how useful risk analysis is, i.e., whether it masks the fact that there is a lot of judgment and consequent uncertainty involved.

Nelkin: I do not agree that it is natural. I was involved in a risk assessment with the Cambridge City Council on recombinant DNA research and they, as a citizens' group, wanted to define risk in very broad terms—in terms of dangers to values and various other things—but they were told that the limits of their mandate was to find risks only in terms of human health. They could not deal with technical aspects because they were not well enough equipped. The definition in these narrow terms is performed by those who have control of the technology. I think that is clear.

Wynne: In answer to John Lathrop, perhaps there is too much at stake to conduct a risk assessment on information technology in the same way that there was in the 1950s for nuclear power. The risks of risk assessment then might have been that people might have chosen to do without it. Risk assessment then perhaps becomes politically possible when it was a case of adaptation in design, etc., within the framework of an established commitment to the overall technology.

Thompson: If risk analysis is broadened to include a cultural analysis of risk then there are such studies on information technology—in fact there is one here in Vienna. The interesting thing about that technology is that it is a bit like solar technology in that it can be all things to all men. That is the reason for looking at it in this sort of cultural framework which can handle different futures and different societies.

What kind of information technology will we have in each future? Will it be a future in which small-scale decentralized computers accompany a developed, decentralized pattern of decision making, or will we have a few large-scale installations and a very centralized form of decision making? The invasion of privacy is seen as a big risk (Big Brother) in the sorts of social contexts that favor decentralization, but as rather less of a risk in the contexts that favor centralization—provided access is constrained by "proper" procedures so that information does not spill out of the "correct" channels.

It is interesting that these are two technologies that share a high degree of cultural flexibility. They can come out very differently depending on the cultural bias they are subjected to. In contrast, nuclear power tends towards just one technology. If you could have a little nuclear reactor at the end of the garden and a man from down the road to come and dispose of the waste from time to time, there would be more flexibility. But, since that is not possible, nuclear power only works for one of those futures and I think that this is a partial explanation, i.e., whether the technology has the flexibility to change cultural direction.

Linnerooth: There are many explanations why risk assessments are currently carried out in the nuclear industry but were not in the 1950s. One explanation concerns the paternalism on the part of the engineers concerned with the technology. They felt they were in control of the risk, and it could be settled by "best engineering judgment". Experts were not asked to make judgments whether a technology was safe, since it was designed to be safe. Many official statements at that time reported that there were no risks from nuclear power. I think there is a similar thing going on in West Germany today.

Douglas: I would like to suggest a sort of quasi-Marxist explanation inspired by Brian Wynne's remark that the risks inherent in information technology were never discussed. This might be because there are few physical risks there. Or it might be that some major interest in protecting that industry from additional harrassment.

There is a lot of correspondence in political journals in the USA about a new class. This new class shows on Alec Lee's diagrams on the distribution of the different sectors in the labor market as part of the shift towards the tertiary or information sector. This sector certainly contains a class often referred to as the knowledge class. The knowledge class is directly involved in telecommunications and information. There is part of its traditional role in teaching and the development of bureaucracy and in the oral communications industry. Then there is the media industry itself, which is large and thriving in most industrial countries.

A Marxist type of explanation would be that the selection of risks for concern corresponded to the class struggle for power. The USA is the supreme case where people are more involved in risk analysis than elsewhere. The argument about risk could be seen as part of a power struggle between the two upper classes in the USA (this is, of course, entirely speculative). One of the two upper classes is the big industrial complex, plus affiliations to it through various occupational links; the other class is the knowledge people who are purveying knowledge in different ways, through universities and through the media. We in the knowledge class are not dependent for our careers or professions on any physical capital so much as upon human capital. This knowledge class is either making a bid for hegemony or trying to maintain an advantage.

On this analysis, risks would be treated as one of the chosen instruments of the knowledge class for attacking the industrialists. It is an effective choice because the media industries, in order to expand, need to reach a larger public, and the largest public is not interested in straightforward information of the encyclopedic kind. It is very interested in scandal and horror, which are well provided by propagating information produced by risk analysis because explosions are horrible and generate plenty of scandal. Thus, following Brian Wynne, the risks of the telecommunications industry itself would come in for some incidental protection.

The class struggle argument would be that risk analysis is our tool (us here today) against others who are much richer and more powerful than us (but who also have to employ some of us as risk analysts on their side in order to counter the attack). Here we are offering our services in this class struggle, which we hardly notice because it is not the classic conflict between the lowest and the middle classes. The argument would be more interesting if the telecommunications industry involved large physical risks, of which I have not heard.

Kasperson: With reference to Brian Wynne's question, I have just finished an article on the social risk of television. I started trying to answer the question of how this technology, which on the face of it appears to be quite hazardous compared with other kinds of technologies, had elicited so little response in the USA. It turns out that there is no shortage of risk assessments, there is a surgeon general's report and a number of other major risk assessments. Most of them, however, turn out to be cultural criticism rather than risk assessment.

Kleindorfer: Are you talking about hazards to mental health or what?

Kasperson: You can look at it as a form of pollution which is essentially hazardous information—there are a variety of social hazards. We include things like violence, stereotyping of various kinds, and just the passivity that is created in individuals. Simply identifying the hazards is interesting; that includes risks to values and institutions, dental cavities, etc. Another interesting point is that culturally these risks are handled very differently. It has a lot to do with how different cultures think about the role and value of television in the society. In the USA there have been many risk assessments but little risk management. This has partly to do with the First Amendment, but it is also very much connected with the political economy, and I think that is the main reason. This does have links with nuclear power in the early stages of the hazard.

Wynne: That would support Mary Douglas' hypothesis.

Kleindorfer: I would like to take up the issue that Mary Douglas raised regarding the institutional aspects of power and the knowledge class. It seems to me a very interesting descriptive task to determine in various policy areas the relative numbers of economists, lawyers and other professions employed in governmental agencies responsible for policy formation. I know something about West Germany and the USA, and there really are very different notions in the ministries in those countries as to what qualifications one should bring as an organization to areas such as health, education and welfare, or transportation.

As a broad generalization, I believe one would find relatively more economists and problem analysts in the USA and relatively more lawyers in West Germany. In the USA, for example, when you look at the people who were involved in the Transportation Department from its inception, there were a lot of economists and cost-benefiters. They probably got there as a part of the surge of belief in the benefits of analysis during the "Great Society" programs in the 1960s. If you look at a number of other governing agencies you will see the same thing, relatively speaking: a very high percentage of policy analysts and economists. If you look at West Germany, there is a very different composition of skills in the corresponding ministries, with relatively more lawyers than in the USA. One of the reasons for this, I suspect, is that in West Germany one looks for the answer to the evaluation process below ministerial level, i.e., one looks for a process solution.

In the USA, we seem to have opted for a more technological Marcusian solution to policy formation. The facts are assumed to be external. One then turns the knowledge class loose on the problem and they come up with the answer. Since they have the best methodology, the answer is indisputable. In West Germany, there seems to be a very different sense of what constitutes the policy formation process and this may be reflected in the very different composition of agency skills there. It would be interesting cross-culturally to consider whether relative composition of skills in ministeries that are focused on risk analysis would reflect cultural differences within a general framework such as that advanced by Mary Douglas and Mike Thompson.

Schütz: I would like to follow what Paul Kleindorfer said on information technology. As I perceive the state of the art in West Germany, there have been analyses on the future consequences of development of information technology and the media. I think this process is not so important. At the moment we have a broad process of discussion on how we should manage future developments on the governmental level as well as on the party level. There are very opposed views between the parties on future media policy and information technology policy and even on the level of the supreme court, which last week rejected a decision on that. I think we are not so much in need of further assessments but more discussions, including the unions who play an important role.

Ronge: The supreme court decision is a very typical and very specific problem. It went around the question of whether we should have private TV or not. This is no problem in the USA, for instance, but it is typical of the German situation. We have public corporations, and TV is looked at as a public entity and a public authority. That raises the question as to whether we should have private corporations and TV programs in addition to that. This is a very specific example. Of course, we have other studies on various aspects of informational problems, such as cable TV problems, and human social problems connected with the computer industry. There are lots of such studies, and we even have a new agency which was established to promote studies of this nature.

Yet, apart from this there is a real cultural difference between West Germany and the USA. In Germany, there is no strong need for this type of study to be integrated into the decision making process. There is a great belief that the experts and the scientists *are* experts. There is rather a severe line between this expertise and the decision making process. I do not know whether this is a problem of the civil service which is quite different in Germany. The mixture between scientific expertise and administrative competence is not so easy in Germany as it is in the USA. Another point, which is again a cultural question, is the belief in formalization and quantification which is much higher in the USA. We do not have that in any field except for nuclear energy, and we imported that from the USA.

Kunreuther: I would like to raise another question. Does anyone know of a case where a risk analysis was undertaken without the particular party feeling that it has to justify its actions? The reason I have raised this question is that in our previous discussions at IIASA of the French system, it was not clear where risk analysis came in. There appears to be an elite group making decisions, many of whom were trained at the écoles polytechniques, have engineering backgrounds, and have been very capable of dealing with detailed risk analysis. But it did not appear as if they were in any way interested in it, or if they were, we really did not obtain any feedback on how it was utilized.

When you mention the case of nuclear power back in the 1950s where risk analysis was not utilized, it was never clear to me if it *had* to be utilized at that time. There was never anyone who was specifically asking that a risk analysis or risk assessment be done. One of the things which seems to be emerging from our discussion is that recent institutional arrangements request that people are hired as cost-benefit analysts because questions are raised as to whether particular hazards are dangerous.

The question that I am raising here was part of the focal point of our previous discussions when we talked about the idea of an acceptable risk level, what is the probability of something happening or not happening. In all of these areas that have been mentioned I am not aware that there has been any explicit consideration of some of these ideas. The interested parties have somehow come together and have made decisions. If decisions are being made in one context where it is not necessary to justify or defend one's actions except to the parties within one's own group, there does not seem to be a great deal of pressure for having any kind of risk analysis.

As a technology emerges, the general public and special interest groups demand certain requirements. The mass media is usually pushing in one direction by saying: "Why are we building these power plants?" The engineers or the people directly concerned with the technology say "We are building them because they are safe and we have figures to prove that they are safe enough." There does not seem to be anything similar for information technology. There is no one asking for an acceptable risk level. That is not part of the agenda defending that technology. There is a different set of questions and they do not appear to be related to risk.

Schoemaker: If you consider risk analysis to be one element of decision analysis, then it seems that people are commonly engaged in risk assessments in order to make choices. However, they may do it solely for the purpose of making a choice themselves. Maybe we should distinguish this sort of individual decision analysis from organizational or societal decision analysis. In answer to your questions, yes, I think it is quite common for risk assessments to be conducted as part of decision analysis.

Wynne: I think there is a distinction to be made there. You are talking about conflicts of interests and what we are talking about are risk assessments that are carried out in order to defend or justify commitments which have sometimes existed for a very long time. My probings around information technology were intended to raise the point about the social role of risk assessment: that either it is performed after a conflict has emerged, for whatever reason, in which case it will never replace political negotiation as a fit response; or it is performed to try to pre-empt conflict and to avoid the need to negotiate. When the stakes are as large as in a big issue like nuclear energy or information technology, no one is going to initiate political and technical debate via risk assessment, until it is too late to be constructive. Risk analysis has, I think, to see itself as encouraging such debate even if so-called decision making "clients" do not want it.

Kasperson: I think that is one path but I think we should not neglect the fact that there is a great deal of routine risk assessment going on, e.g., if a chemical industry develops a new type of pesticide, a risk assessment will have been carried out many years beforehand, and the same if a new automobile enters the market. These are not formal risk assessments, but an enormous amount of quantitative risk assessment has gone into that new product. There is an awful lot of built-in risk assessment that goes on that is largely invisible.

Session II

- o What heuristics do interested parties utilize?
  - — Uncertainty avoidance
  - — Resolution of conflict
- o What impact do systematic biases have on decision processes of interested parties?
  - — Importance of concrete data—salience
  - — Probability biases
- o What impact do different policy instruments have on decision processes and welfare?
  - — Market mechanisms
  - — Information
  - — Incentive systems
  - — Regulations
- o What is the role of the political process in shaping these arrangements?
- o What is the impact of different cultural settings in shaping attitudes and perceptions?
- o What are the advantages and limitations of different arrangements as they affect:
  - — Power of interested parties?
  - — Responsibility of parties?
  - — Knowledge and information?
  - — Credibility of institutions?

Lathrop: Just what is it that we are trying to address here? Is it designing better negotiation procedures or designing better analyses to feed into those procedures? There seems to be strong disagreement on this point that requires clarification. The point of my talk was that the~e ought to be a role for utility functions and analyses. But then Howard Raiffa commented that the essence of the problem is resolution of social conflict. Should we talk about how analyses should be shaped to serve best the resolution of social conflict, or should we talk about how procedures should be shaped to be conducive to the resolution of conflict?

Raiffa: I feel very strongly that conflict is at the heart of many resolution problems and I am a firm believer in very strong democratic political processes to resolve these problems. I happen to believe that we do a miserable job in conflict resolution and I believe that analysis could be instrumental in helping individual negotiators and the process itself in arriving at better resolutions; better in the sense of squeezing joint gains and making all parties happier. I think on the whole we have a very adversarial society, not only in the USA, but also from nation to nation, and there are many possibilities for both sides to win. We do a miserable job and analysis has a role to partially help all the individuals and interveners in this process.

Kunreuther: I would like to pursue two aspects of Howard Raiffa's point. First, why is it we do not do a good job and second how can we deal with it? I would like to leave it fairly open at the moment in terms of the whys as I think that relates to different problem types, different cultures and international settings—in fact, a number of these themes have been instrumental in bringing this group here today.

Schoemaker: The issue of conflict, being very real and present, might be dissected into dimensions along which people have conflicts. Value conflicts could be one, perceptional conflicts about how the world really functions could be another. What are the different roles of analyses along these different dimensions?

Raiffa: I have a feeling that they are both present, and its good that they are, because as a rule of thumb the more dimensions of conflict there are, and the more people with widely different points of view, the more these differences can be exploited in obtaining compromises. The secret of obtaining compromises is diversity of opinions and that is where analysis has to play a role. I think a large part of the role of analysis is to seek out differences in value trade-offs in issues in order to come to agreement.

Nelkin: Let me add something that I think may clarify: The role of analysis is not necessarily to achieve consensus, but to achieve a means of compromise—there is a difference here which I think is important.

Raiffa: Absolutely, I agree with you.

Nelkin: Often analysis is thought of as a means of consensus because it is expert opinion.

Raiffa: That is more of a semantic question; what you really want to do is to get society to adopt different solutions or different choices.

Nelkin: To make a choice.

Raiffa: And that is compromise. The point is that you make one choice now, but if you could find other choices that are different, that are more creative and will improve the lot for everyone, then that is what we want.

Kleindorfer: Research on small group behavior suggests certain aspects of choice situations which inhibit consensus and compromise solutions. Such research would certainly suggest that one of these inhibiting factors is complexity. For example, the reason people select fixed alternatives in the first place is because they at least know what they are and what they are about. The problems of actually generating additional ones and evaluating them with the same amount of certainty that one has about scenarios one already knows is a troublesome process. I hear this issue of the complexity side and the generation side being brought up continually and it seems that it might be very fruitful to consider it as a possible area where decision support, scenario generation, or other process-oriented approaches can help sort out the issues.

Schoemaker: I would like to ask a question, possibly to Howard Raiffa, which follows from that. Let me just take an extreme position in order to make the point. On the value side it is hard to say that some values are right and others are wrong, so compromise seems the way to go. On the perception side, however, I am not convinced that this is the right approach. Although we should allow participation by all sciences on the fact side, there is, in an ideal sense, only one reality. Why should we follow a compromise strategy when we are dealing with perceptions about the world itself. For instance, one group may have one set of predictions as to what will happen in the future and another group has another set. If we assume both to be equally correct, or simply acknowledge their right to disagree, then compromise is the result. Would it not be better to strive not for compromise, but synthesis, in the sense that we seek one science and not multiple sciences. How should conflicts along the science dimension be resolved?

Kunreuther: Let us open that question to a broader section of people because I think there may be some debate on the point about *one reality*. I am sure there are people who have some feelings about how we deal with the perception side, where each of the groups may have their own view of the world. We heard from Elizabeth Drake who was looking at the risk analysis part, and showed us that this might be a problem; Chris Mandl also brought that up in our LEG study on very specific kinds of data where there were different views of the world. How do we deal with this type of problem?

Fairley: Just a quick comment on compromise and synthesis in adversarial situations in response to Paul Schoemaker's remarks. It is sometimes suggested that the quantitative results of two opposing analyses presented in a hearing testimony be averaged together. Such an average often represents a compromise that represents no coherent estimate of any well defined, unknown quantity and is therefore literally nonsensical. A synthesis by contrast might use insights from two analyses as appropriate to determine a third estimate of a well defined quantity. The numerical result might, however, be in between the results from the initial opposing analyses.

Schoemaker: If you have an adversarial system in science you must recognize that it is dialectical. One should not be happy, in my view, with having contradictory sciences. Consequently, one should not be happy with compromises, where theory A is deemed legitimate as well as theory B and then bargain, but instead one should search for a synthesis.

Douglas: If parties in conflict try to discuss their different perceptions of the world, surely Paul Schoemaker must be right, there will never be satisfactory compromise. It is possible to find a common position by avoiding the assumptions each party makes about the world; they could insist on coming to agreement about procedures for establishing common values.

Schoemaker: The rules of evidence and the process of judgment.

Douglas: Yes.

Kleindorfer: Well, I do not know because sometimes the rules of evidence are very different according to the situation. If you are arguing for something where relevant data is available, the rules of evidence might be argued within one school of thought, whereas if you are arguing in a situation where you do not have enough data for this purpose, then other rules of evidence might apply.

Kasperson: What does one do about the problem that not everyone who comes to the bargaining table comes with the same resources? There is an enormous disparity in technical and financial resources, etc., and how should that problem be dealt with in what we are talking about?

Kunreuther: In terms of looking for compromise, you have different starting positions, as it were. What does compromise mean in that context when you are trying to reach a solution where there are a number of interested parties?

Kasperson: Even how they develop conversation in that context.

Nelkin: Or expertise.

Raiffa: I think it would be nice if there could be a group of analysts at universities, etc., not only supporting the needs of the industrialists but also serving the public need. How do you get groups that will do work for environmentalists, consumer interest groups, etc.? There are industrialists who believe that their cause can be best served if there was better analysis by environmentalists and consumer interest groups and they have offered to subsidize analysis by those groups but those offers have been rejected because the position would not be credible. That is one of the questions: How do you get funding for analysis that is credible for these types of groups? In the USA we are very bad in this respect.

Nelkin: There are a number of models around the world that have been developed. The Science Shops in Holland are one of the most interesting. They are designed explicitly to distribute expertise to groups involved in negotiations of various sorts who do not have access to it normally. There are a lot of similar efforts elsehwere as it is an area of ongoing experimentation, particularly, in risk disputes.

Kunreuther: How well has it worked in terms of its ability to resolve conflict or perhaps cause more conflict?

Nelkin: In many cases it causes more conflict because the issues are basically not technical but rather are concerned with resolution of conflicting values. Public interest expert groups, in the course of their discussions, often reveal or highlight value conflicts. But they may also help resolve disputes by making people feel more comfortable with bureaucracy. If they have some sense of equity in decision making, this gives them a sense of participation in the decision making process. In a statistical way I cannot attest to the extent of the success of such groups. There have been lots of experiments in other countries—the Burgerdialogue in Germany, the Berger Commission of Canada, and various experiments in Swedish circles which are designed parimarily to obtain a better distribution of expertise.

Salz: My view of the Dutch Science Shops is that they have been rather successful.

Kunreuther: How have they done it? Could you give us an idea of how that has worked? Are they in the spirit of the question which Howard Raiffa carved out, i.e., we want to get analysis into the process so that we can do things better. Should we provide resources to give each of the different groups some opportunity to do analyses as Roger Kasperson asked? Have these groups done analyses or what actually happens?

Nelkin: They do analyses, they do it free. Basically it works on a volunteer basis with some money provided by the government which pays for xeroxing, etc. I think they now have some foundation support basically from the science council via the ministry, but this has been controversial as some of the Science Shops are quite radical. It is a decentralized movement organized in six or seven university centers throughout the country. They are separate and distinct centers, often with different ideological biases. Some are reluctant to take money from the science councils or the ministry of science and just have university funds, but mostly support comes from "release time" of faculty members from the universities. Their biggest problem is alerting the public to their availability. They have more trouble getting users—and they will not supply information to just anybody. Essentially what they do is research for unions, environmentalists, or neighborhood groups who need to support their positions, i.e., they do a lot of work for people who do not have access to other kinds of funds. For example, a common neighborhhod problem is flooded basements because the level of dykes is not regulated properly. This involves technical arguments that are not terribly complicated, but there are some engineers who specialize in this and they will act as liaison officers in various negotiations.

Schoemaker: They have them in the US now too.

Nelkin: Yet, but they have been defunded.

Schoemaker: There seems to have been an increase in organized student participation on such issues. What is the level of participation in Holland?

Nelkin: The most active are very young, but they do have a number of older people—I do not know the exact age structure in the Science Shops. It partly goes along with political predilection; generally people with a somewhat radical bent are involved.

Schoemaker: But they engage in more than "objective" analysis; they also take on value issues. They do not always make a clear distinction between values and science.

Nelkin: A lot of work is synthesis, not scientific research. It involves somebody who will volunteer to bring together the existing state of knowledge in a given area to bear on a particular problem. In other cases people do mini-experiments. They will then advise on tactics, on how to use expertise in a negotiating process, and also on factual material.

Douglas: Do they go home in the winter? They sound like mercenary armies in Europe who used to go home in the winter and had various conventions for reducing bloodshed and generally keeping warfare in bounds.

Nelkin: The Science for the Citizens Program of the National Science Foundation was in part modeled after that.

Schoemaker: Which is also now defunded.

Nelkin: It is an interesting example of a possible way to distribute expertise, but there are others around dedicated to the same purpose.

Kunreuther: I have a sense that what you are saying is that it would be nice to see analysis brought in. We see an example in the Dutch scene where it has worked well. There may also be examples in other countries which might be useful for us to try and explore because if we can begin to get some prototype models where this has worked well and find out why, we may be able to get to a point where we can think about how we can resolve some of the questions.

Nelkin: Let me start with a warning because I have been looking at examples comparatively. I agree with Mary Douglas when she suggests that these models are not necessarily transferable from one country to another; that they are really culturally based. If you look carefully at the Dutch case and the kind of assumptions that underlie it, there are reasons why it assumes that form there but not in other countries. You have really to look at the cultural and political expectations of the Dutch, based on historical circumstances, to understand the shape of its participatory experiments. So too in other countries.

Kunreuther: Would you talk for a minute about why you think the Dutch experiment worked well there? We can then maybe turn to some other experiences where it has or has not worked well so that we can begin to get a spectrum of thoughts on that.

Kleindorfer: While you are responding to that, what does it mean to work well?

Nelkin: Why has this experiment developed and persisted in Holland and even obtained governmental support? This might sound very simplistic, but Holland is a country which has traditionally experienced conflict because of its religious splits and also historical fragmentation. Conflict resolution has been terribly important in

Holland for a long time and there is a tolerant acceptance of conflict; it is not rejected as necessarily negative. Consensus is somehow not talked about in the same terms in Holland as elsewhere. As a country it is interesting to study with respect to conflict and conflict resolution because this is so fundamental to its own history.

Schoemaker: That is my impression too. Moreover, in Holland students are heavily subsidized by the government. With tuition fees very low, some may feel a real obligation to repay society. I do not see such an attitude in the USA, where someone may pay $7000 per year for tuition. This contextual difference maybe of significance in understanding differences in student attitudes across cultures.

Douglas: This leads me to a crucial question. The fact that the government feels disposed to subsidize students is not sufficient to settle the nagging worry that we have had in several papers that risk analysis, when there are very deep differences among the stakeholders in conflict, only serves as a form of massage to the most powerful—who are going to win anyway. Therefore, the real question is whether risk analysis really works well or whether it is just decorative, as Volker Ronge was suggesting, by standing outside the real issues, or, according to Kasperson's law, not standing outside them, but simply making no difference.

Nelkin: I am disturbed because working well can mean a million things. Risk analysis could work well because it involves a short-term dispute but it allows people to accept a technology that is not necessarily positive, but that could be a definition of working well.

Douglas: If risk analysis enabled a decision to be reached, that could be a good criteria for working well.

Schoemaker: The difficult question here is what criteria to use in assessing whether decision analysis has worked well?

Kleindorfer: Working well might be understood in some logical *ex ante* expected pay-off sense. But equally important in the political arena is who is accepting responsibility for the results of planning, whether analytical or not. In the Three Mile Island situation described by Roger Kasperson, there was something like ongoing social consensus about the utility, etc., of nuclear power. But when things blew, everyone brought up the fact that a warning had been issued *ex ante* and nobody was willing to take the responsibility *ex post*. Thus, both *ex ante* and *ex post* felicity seem to be important. *Ex ante* one wishes informed consensus, so that there is something linking the outcome and responsibility for it to a common understanding that the people have for the problem. Perhaps, as Howard Raiffa suggests, analysis can contribute here. But there are also many problems in balancing consensus and responsibility. If one opens a sensitive public issue up to public participation and analysis, there is a high probability that you are not going to get consensus *ex ante*, nor any sense of social responsibility *ex post* if something goes wrong. Thus public participation and analysis may promote *ex post* felicity, but it may also hinder *ex ante* consensus and feelings of collective responsibility *ex post*.

Nelkin: Most people who study participation get around that dilemma by saying that there are other virtues in participation besides the actual decision, and that there are real costs in nonparticipation in terms of declining trust in government, and in the feeling of being a citizen.

Kunreuther: There is one question I would like to put, in the light of this. One says that this works well, that all the parties go home and say that they are delighted to have been part of the process. We may not agree with the outcome but we can live with it. That is one kind of a solution where one can say there is a good feeling at the end of all this. Then there is the other issue that Paul Kleindorfer raises that may be more difficult. Even if we are not delighted with the outcome, we now all feel part of the process and we are going to share in the consequences.

I heard Paul Schoemaker suggesting that in Dutch society there may actually be this feeling of responsibility for one's actions afterwards and a willingness to accept the solution as part of their life—in a different way perhaps than one would have had in other societies.

Kasperson: I should like to suggest a second process; there are some people here who know far better than myself so I will be brief. The remiss system in Sweden is a very intricate system and when risk analysis has been done for some particular subject it is then systematically farmed out to a number of large organized interests in that country, all of whom do their own fairly careful analyses of the issues associated with that document. There are mechanisms for ensuring that subsequent analysis which is adopted when a decision is made by a minister, or whoever, that the decision is made on the basis of the original document and the remiss process. The remiss process is very valuable because it not only brings in the diversity of analysis that we were so worried about, but there is also a very strong consensus-building aspect. Once a decision has been made, of course, it could be rejected. But you feel that you have had a fair try, and there is then some responsibility for making the policy work.

Nelkin: It is very interesting to compare the remiss system with another system in Holland where "policy intentions" are circulated. The Swedish system is based on a sort of cultural monolithic consensus whereas the Dutch system is based on a lot of expected conflict. This provides a very good contrast because the kinds of structures that have been developed to resolve conflict come out quite differently in those two countries, since they are based on two different cultural models.

Kunreuther: Do they use analysis in very different ways in terms of defending their positions, in terms of presenting their arguments when they know that consensus is what they are striving for? Do we know much about that—is this a research issue for the future or is this something we know? If you were going to write a comparative paper about the Swedish and Dutch systems would you be able to say something about where analysis came into the picture and how it was actually used as part of the decision process?

Nelkin: I would be able to say something about how analysis is conveyed to the public, but I would not be able to say much about the technical process of analysis. I think that the key variable is not what the experts do, but who the experts communicate with, and how they use that as an input into the decision-making process.

Drake: As somebody who has done both worthwhile and useless risk analyses and has seen them influence or not influence decisions, I sort of cast this discussion into a research need. It seems to me that we have probably too broad a spectrum of cultures and issues to be able to come up with one approach that will be universally applicable. I think each country is gradually interested in improving its own situation, whatever the societal goals are. It might be useful to do some fairly detailed case studies across technologies and cultures to look at controversies in which risk is apparently one of the decision factors, and then look at whether risk analysis was used, whether it was overkill for the problem, whether it was useful, whether it was used to send up a smoke screen for the real issues, or whether it was just totally irrelevant. I think it might be useful to understand where analysis might be better used in the future.

Nelkin: I was too embarrassed to ask before, but it turns out that many of us have the same problem—was is meant here by formal analysis?

Kunreuther: One assumes we all have our definitions straight and although I think Howard Raiffa started off by saying that analysis could be defined in many different ways, we could probably do the same for formal analysis. I am not going to try and answer this question and would like Howard Raiffa to have the initial say as to what he thinks formal analysis entails.

Raiffa: I define it in a very broad sense. It involves systematic, conscious thinking. It involves the art of taking a complex problem and decomposing it into different parts. It does not necessarily have to involve quantification or mathematics. I think that as analysis becomes more intricate, one has to use more tools, and mathematics is a natural tool. But I also think that "a little analysis goes a long way".

Schoemaker: What is formal about it?

Raiffa: Conscious, systematic—it is an attempt to document thinking so that others can understand the process, peers can review it, and contributions can come from different experts. That is the nature of formal analysis.

Nelkin: But in this discussion it has taken on more of the formal role. I started out with that kind of assumption but in the course of discussions over the last few days the word "formal" has become more important and it has taken on some kind of a structure which appears to be much more rigid, at least in the discussions here.

Kunreuther: Incidentally, we had a difficult time in phrasing this question. We were not sure whether we should say risk analysis, quantitative analysis or formal analysis. We finally focused on formal analysis because we wanted people to think of the process as being structured. Perhaps we could have used the word "structured" instead of "formal".

Kleindorfer: I think that formal has the connotation that Howard Raiffa properly underlines, namely that formal analysis is transferable within a particular discipline so that one really has to understand the sociology of that particular discipline to know what would be acceptable as formal analysis.

Raiffa: As an aside, this Institute is called the International Institute for Applied Systems Analysis. But people asked what this means. The reason why this title was accepted by 12 nations in 1973 was because nobody could pinpoint its meaning. When people ask what I mean by systems analysis I prefer merely to change the word "systems" to "systematic". The question then becomes: What do we mean by systematic analysis?

Kunreuther: But what I have heard both Roger Kasperson and Dorothy Nelkin saying without defining their terms is that analysis is used in these countries. This suggests that there is some kind of systematic analysis or some kind or procedure that is used. What Elizabeth Drake is saying is that maybe we had better do some research to find out and define in a little more detail what we mean by the word analysis. It tends to be a little confusing when you begin to look at the institutional structures in different countries and the way each of these different groups look at a particular problem. You cannot necessarily transfer an analysis from Holland and bring it into the USA because there are too many differences between the two countries.

Nelkin: Why did you not use systematic in phrasing the question?

Kunreuther: We could have. But I do not think we would have changed the tone of this discussion. Do other people think it would have changed the question?

Larichev: It would really change. I think if you put the word systematic instead of formal analysis, it would be very different.

Linnerooth: In our discussions, we thought of having "risk" in the terminology, but that was too narrow because of the different problems "risk" encompasses. We decided to change the term to quantitative, but that was also too restrictive, so we changed it to formal. If we had carried on we might have decided on systematic or even something else.

Nelkin: I have nothing against quantitative analysis, but I think that there are other forms of analysis that are useful in these areas. There are areas where you cannot use quantitative analysis so you really want to incorporate both of them into your question.

Lathrop: Coming from a physics background I would like to see us characterize the ideal form of the analyses we are talking about: not as systematic, quantitative or formal, but as objective analyses. I do not mean objective in the narrow sense, but in the broad sense as defined by repeatability. That is, the analyses we are talking about are those parts of the process that should not vary as a function of the agencies or consulting firms performing them.

Schoemaker: I would like to raise an issue that is related to this. With formal analysis one has to make the rules of the game explicit to oneself as well as to others. The approach is reductionistic in that one separates values from beliefs (i.e., predictions). You allow other people to challenge you in certain domains. For example, you might be willing to change your beliefs without affecting the rules of evidence as you laid them down. Now there may be people who object to this approach precisely because you allow yourself to be open to scrutiny. The moment you commit yourself to an explicit decision procedure, you open yourself up to the possibility of being challenged and having to change within your own world of consistency. For example, I have encountered people who prefer not to get into a formalized thinking mode because they are afraid of where it might lead them. Specifically, when new information is introduced, the formal mode may force a revision of opinion, which may be disliked or not preferred.

Kunreuther: I think Brian Wynne made that point in our earlier discussion; i.e., the fact that only certain people would want or get into the process. I would like to see a little elaboration on that because I think we got into this discussion with Paul Schoemaker's comments on various types of procedures. The point raised as to how one can get into that if some groups refuse to follow the rules of the game and hence would not put their cards on the table. We have found that to be the case in our LEG case studies; it is very difficult to find out explicity what is happening.

Wynne: I would like to go back to previous comments about the specification of goals as part of the process of eliciting utility functions. I used an example where people rejected cost-benefit analysis of an explicit and systematic kind, i.e., where we say: You are making decisions implicitly every day, why do you not let us do it for you systematically? We could point out the consistencies of its use in one place in comparison with others. My interpretation of that rejection, and this is a sociological interpretation, is that they want to retain control of the situation and want autonomy. They do not know who you are, out there with your systematic approaches and your rationalities and your demand to separate values from facts—you are like people from Mars. If they specify their goals or utilities too precisely for you the analyst, they have enough sense to fear that those expressions of themselves and their hopes and values will be used too simplistically and too rigidly. They lose the chance to revise their values and objectives in their own "social space". They are being asked to give up even more of their lives to alien beings—why should they? I do not mean that rhetorically; you as analysts and "decision makers" are a different society, and these things are coming down through the cloud of the unknown and depositing themselves like meteorites or something. We have to take this seriously. It is not just something we can moralistically dismiss or treat rhetorically.

Kunreuther: I want to push Brian Wynne a little on this point because I think that we are all looking for ways of doing things better. Groups may come into the process with hidden agendas and it may very well be that these groups are saying: "You are a foreigner, we do not understand where you come from." Alternatively, they may simply be saying that because they do not want to be part of the process, even though they fully understand exactly where the other groups come from. There is a real difference at this point in terms of how you begin to prescribe. In other words, we have to have some understanding of what the role of these groups is.

Consider the scenario where people understand where you come from but refuse to play the game. In other words, they say they do not want to do formal analysis because, as Paul Kleindorfer was suggesting, they do not want to lose control of the situation. Is that something we just accept as part of the decision process, or is there some way in which we can move in the direction of what Roger Kasperson and Dorothy Nelkin were pointing out for Sweden and Holland?

Wynne: That is a question firstly about whether those cases were successful, and secondly whether you see the role of risk assessment as eliciting the underlying motivations, interests, values, etc., that people possess, and thus expanding the terms of social debate. Is the process of risk assessment of a formalized kind, helping to elicit those authentic commitments underlying whatever is said in political or technical vocabulary, or is it suppressing them? We then get into the problem of ambiguity and the extent to which we feel we should retain our ambiguous concepts or should systematise them into absolutely locked-in precise meanings. It may be that keeping hidden agendas hidden is a good idea, in that it is perhaps helping us to avoid deeper conflicts which might even be quite destructive if they were explicitly laid out. I do not see any ways of producing some generalized statements about that. You can only go from one empirical case to another in the hope of insights that allow generalization. The problem is to know how we do or should move from one level of conflict to another in the political and social process, because I do not think you can intrinsically define the issues as belonging to one level or another. In other words, how does explicit conflict arise and change in connection with societal issues, whether it is LEG or some other issue, and how, if ever, does it go away? I have never seen it go away. What we use as criteria for successful risk assessment depends upon what we think is the proper social role for risk assessment. Let us discuss that too.

Kunreuther: But what we are coming up with are examples of success in one form or another. I feel that the group is saying that we have to do things better. We can pose these dilemmas but we know that somehow we have to come up with solutions. We have heard some cases where things seem to have worked well.

Kasperson: Let me try to answer this. A good example which was successful in providing useful analysis, and maybe Jim Dooley can comment on this, was the Berger report in Canada*, where researchers went out into villages and listened very carefully to the kinds of concerns that people had and the different ways that the new technological development could affect their villages. Nobody forced *them* to do an analysis but someone had to do an analysis to cut through all the information to see what it all meant. That seems to me an example of a process that aided the policy process in a productive way. People were not forced to do an analysis but still obtained the kind of information that was extraordinarily valuable for the policy process.

Douglas: I would like to have some guidance as to how much this form of analysis is normally geared to the less-privileged members of the community. There are large numbers of people exposed to hazards, but they are mostly the underprivileged who are actually more vulnerable to dangers and also less able to move away from them or to get compensation. When I was listening to Dorothy Nelkin talking about the Dutch Science Shops, I was trying to imagine what they were really like. In a sense I can guess fairly well who comes to them and who are the people who are being served by them. I can well imagine that the people we should be concerned about are not actually part of the debate.

---

*Berger, Thomas R. (1977) *Northern Frontier, Northern Homeland: The Report of the Mackenzie Valley Pipeline Inquiry*. Vol. 1. Ottawa: Ministry of Supply and Services.

Nelkin: That is a problem that the people in the Science Shops worry about. They do not reach a lot of people they want to reach because the people who usually seek such information tend to be middle class, that is, those who know how to get it.

Douglas: But they might be proposing themselves as representatives of the inarticulate people who might have very different cultural perspectives.

Nelkin: They do not want to go out and openly solicit because they have this concern about manipulating people. To what extent do you serve as a spokesman for groups and how far do you go in transferring your own values to these groups? They are in that kind of a dilemma.

Thompson: In Cowdenbeath, a city involved in the Mossmorran Bay siting situation, there were some people who were in a poverty-stricken situation because of high unemployment. There was a recent television program where the interviewer asked the townspeople what they felt about the new installation. They all wanted the installation because they were used to working as miners and were not worried by risks: they simply wanted to work. Apparently, the misinformation there is once the installation is built, they will find that there is no work. They were very keen to accept risks and did not agree that the risks were as high as what people were saying, and even if they were that high, they were still prepared to accept them.

Raiffa: When our committee [CORADM] was first formed, we each introduced ourselves and we tried to be a little searching about our hidden biases. I admitted to a bias of using the method of an analyst, which is one of the statements I have heard mentioned here. I am personally concerned about how the analyst can protect the interests of unorganized, statistically anonymous individuals. Those people who are readily identifiable and are harmed or helped by policies can usually organize themselves. Those who do not organize themselves are the people who are affected indirectly by the effects of the decision. Policy workers are not accountable for the secondary or tertiary indirect effects of their actions. That is where the analyst is in a position to help.

Kunreuther: One aspect which we have not explored fully is the equity/efficiency trade-off. From an efficiency viewpoint one could argue that anyone who was really in a position to be harmed will either get their resources together, or will think the situation is not important enough to worry about. The equity that Mary Douglas and Howard Raiffa are bringing up is related to how you transfer resources to some of these groups who are being hurt because they do not have resources to fight for their cause. Another aspect of the problem which cuts across both the efficiency/equity trade-offs is the information processing ability of individuals and groups to process information. We found, from our LEG studies and in reading Lee Davies' book *Frozen Fire* (published by Friends of the Earth), that many groups of people who were at a disadvantage and who were directly affected by an LNG tank or terminal, were not aware of the hazards they were facing. Therefore you have this real problem of whether the experts or people who are informationally rich should make other groups aware of the situation.

Wynne: The people who build the plants are not always aware either, and in any case they will not get others interested except by telling them there is a risk, which is a risk they do not like.

Kunreuther: The reason I am pointing it out is because they are often made aware of it after an accident occurs. They have not been made aware at the time the plant is being built. In fact there is no incentive for interest groups to do anything that might rock the boat if they want to get things through in a reasonable period of time. The dilemma appears to me to be between whether to provide all the interested parties with detailed information and perhaps not build new plants, or not to give everyone this information because it might hinder "progress". How do you deal with this problem?

Thompson: But do people want information? Some people do, some people do not. Some people are forced to have information they do not want and yet get cannot get the information they do want. The first question is, do they want it?

Wynne: To support Mike Thompson, I would surmize that you might be able to get a group to make an abstract statement of the risk involved in the proposed installation of, say, a power plant which means that the information is there available to everyone. I imagine the result would be that nobody would take any notice at all. I think the only way of getting the information to stick is when it becomes meaningful: when you are actually going to open the plant.

Kunreuther: But it is too late then.

Wynne: Of course it is, and that is the problem; but it should be recognized as a fundamental point about the social nature of knowledge.

Kunreuther: I would like to refer to Mary Douglas' question of helping the downtrodden. I would put downtrodden into the category of those who are informationally poor. When you are information poor you are downtrodden as well as poverty-sticken because you do not know what to do. The question Mike Thompson is raising is, do they want to be helped? What types of value judgments are implied by the actions one takes?

Kasperson: I think it is really an ethical question. It concerns the ethics of risk imposition and what kinds of responsibilities you take on when you impose risks on people. I agree with what Michael Thompson says and I think that many people do not want that kind of information. That is fine because that person has made his own judgment and the moment you start making that judgment for that person then it seems to me that you are really violating his freedom. Moreover, if you are serious about achieving consensus to the maximum extent possible, it is an idealized state in itself, then anyone who is involved in imposing risks on individuals or communities has the obligation to obtain the maximum degree feasible. It is not only provision of information, but it is also an attempt to make sure that the person has understood the issue.

Schoemaker: Would that not be the role and responsibility of the government?

Kasperson: If you are serious about doing it then the developer should get someone independently to provide that information. It is a very difficult problem and there are all the difficulties of different levels of understanding. We know a lot about public participation and the difficulties involved, but you need to begin with a clear definition from the start of what the ultimate responsibilities are and try to build from that a respectable process.

Drake: I think Mary Douglas has raised an important issue and I do not think that we are going to solve it here. If you look historically at the past siting of industrial facilities, you will find that usually they are clustered in one area and in an area that is surrounded by working-class houses, or light industry. That is the way society used to function. People with money did not want to live near one of the industrial areas. It is only recently that people have become concerned with the risks associated with technologies and that the issue of remote siting has emerged. Suddenly, industrial facilities are being installed in remote areas and these are the areas where the wealthy live and play. I think Michael Thompson's example of Mossmorran is good because the people in Cowdenbeath perceive the risk to be low. They have lived with higher risks, they perceive that they may have some some jobs directly or indirectly as a result of that project. On the other hand, the people in Dalgetty Bay, which is primarily a nice suburb of Edinburgh, are very upset and are taking a lead in opposing the project.

Thompson: The risks today are also much greater.

Drake: To some extent yes, but what we are really saying is that things are more risky for the middle class who traditionally avoided industrialized areas.

Douglas: But I suspect the risk analysts and decision analysts are taking on the responsibilities of governments. This function of actually knowing what locals feel was a kind of collective responsibility in the old days.

Kleindorfer: What has changed from the old days is that we now have such an array of risks that the government is talking in the name of the public, so that the process of informing the public is a costly one, not just from the government's point of view but from that of the public too. This raises a difficult cost-benefit issue—how much money is one willing to spend to inform the public?

Wynne: I think we are in danger of doing what many participants do at such meetings, which is to trade symbols across the heads of the working classes without really getting to grips with the real issue. On the one hand we are praising their rationality, and on the other we are damning their irrationality. There is a constant lament that people will only get up off their seats and take an interest and get stuck in when it is too late, which is when the siting question arises, and inevitably debate then tends to be destructive. We ought not to forget, and it has been a repeated analysis, that educational systems such as the British one, are organized on the basis of stratification. Elites are at the top of the system and are trained and encouraged to think in abstract, general, universalistic terms, whereas the masses are churned out at the bottom, turn back into their folds, and are encouraged to think in informalized, concrete, particularistic ways. Then, we go along and blame them when they will not think in universalistic, abstract terms so we start throwing universalistic information at them about risks to site a facility at one place or another.

On the one hand, the system that we actually profit from, live comfortably within, and act at ease with is telling them to do one thing via their education and socialization, and then every time there is an issue, a big road show comes along and tells them to do something else. It might be a shrewd means of keeping the vast majority of people silent and disoriented, but I do not think risk assessment should be part of that. Why should we allow ourselves to get away with that anyway? I think that this relates to the point that Roger Kasperson mentioned about the Berger Inquiry, and I think it is one of the interesting things about that inquiry that Berger did go along deliberately and ask the Indians about their identifiable social and cultural aspirations, not just about a national technological risk associated with the pipeline. The whole of the Indian economy and their ideas about nature, their view of their destiny in the most spiritual sense was incorporated into the agenda, to be elaborated in a serious manner, and not just as a matter of exotic entertainment.

Nelkin: It was specifically stated that this kind of information was to carry equal weight with technical information.

Dooley: I have one comment on the Berger Commission. First of all it was an inqiry that really did look at native rights of people who were not technical people, in effect, nomads. In this instance, what we are calling analysis was completely discredited. To illustrate, a pipeline company went up into the permaforst region, which is a very difficult, ecologically sensitive substance. In permafrost, if you drive a truck over it you can start off a series of thaws which each year thaw more, and you end up with vast gulleys of marshland in no time at all. A good deal of the land where the pipeline was to run was through permafrost. The pipeline organization said they would experiment, and did so, but excluded the public from observing the site. The site was dismantled and it was claimed that pipelines do not harm permafrost. If that is analysis then it is the kind that can really discredit. In the Berger Commission, analysis played a very small part.

Another point which is related to other points mentioned here is that one of the leading reporters of the *Global Mail*, one of Canada's most influential papers, went to the inquiry with a typical establishment point of view to report on the inquiry. The reporter returned and wrote an impassioned book supporting native rights and went on to say that when he heard the native people talking about how they perceived this project, and their position on it, it had changed his entire view*. I point this out becuase it has been said that people tend to stick rigidly to a framework but I think this is a very unique inquiry and it broke new ground in demonstrating sincere concern for minority issues.

Kasperson: Before we become too carried away with the possibilities of the alternatives, we should recognize that the kinds of problems we are talking about only work with certain institutions. I really believe that the outcome must be kept open yet most institutions are not ready to believe this. It requires a high level of commitment to the values associated with the democratic processes. What it says essentially is that if the values that you desire for democratic process are sufficiently important you may forego the best location or it may take you five years longer to get your LEG sited. So it is very difficult. We are all aware that we cannot just pass these problems off to the existing institutional structure because that institutional structure itself is a political outcome and we are talking about changing that political outcome. So incorporating the analysis we are talking about is really not possible without institutional change.

Drake: I think we are being very elitist in talking about representing the true information to the informationally downtrodden. Who is going to decide what is true? You have a complex set of issues and there is no unbiased advisory committee. The government agencies have advisory committees, but, for example, the US Department of Energy is developing coal gasification technology but they do not let EPA make environmental studies of their pilot plant effluents because they do not want to know about potential environmental problems until the technology is proven. Our whole society is riddled with biased interest groups. The Berger Inquiry was an exception where we have one well intentioned individual and something good came out of it. But I do not think that that is assurance that we are going to find many such people.

Kunreuther: I would like to bring the discussion to a close. At the beginning of the meeting we asked ourselves if there would be a chance for some synthesis and we have tomorrow for that. After this discussion I would certainly say we are not information poor, but information rich. I will briefly summarize the ideas that have come up during the course of the discussions without trying to make any attempt to synthesize these issues.

One of the points we started with was the fact that conflict resolution was a critical aspect since we had many different parties who had different views of the problem. Conflict resolution could have been at the value end or the perceptual end but there was a feeling that it would be good to find a compromise but not necessarily consensus. One way to begin to do this is to focus on the rules of evidence and in some way judge low-probability events, but this is difficult because low-probability events do not have a detailed statistical database on which to make these judgments. It is also difficult because we have many interested parties some of whom have access to financial resources; others of whom do not. This immediately provoked a discussion as to whether or not we should give the "have nots" access to resources and if so, how much.

The next part of the discussion involved examples of the Dutch Science Shops and how they work. We also had examples of how the remiss system worked well in Sweden and we also heard about the Berger Commission and how that worked well in Canada. What emerged was a feeling that we do have examples where analysis has been part of the process. We have not fully defined what analysis is, although we

*O'Malley, Martin (1976) *The Past and Future Land: An Account of the Berger Inquiry into the Mackenzie Valley Pipeline*. Toronto: Peter Martin Associates Ltd.

came up with a definition of conscious, systematic thinking which could begin to help us think about where this could help play a role.

We ended up with a dilemma of how we were going to judge where analysis could be useful. In particular, there was a question of equity issues: equity not just in terms of the resource base in an income or wealth sense, but also in the information sense. It never became clear to me whether it would be a good thing to provide people with a lot of information because that might produce more conflict. Whether that is good or bad depends on the outcome, i.e., whether or not compromise may actually be enriched by the extra conflict produced. I therefore turn to one of the research needs that Elizabeth Drake suggested: to try and deal with some case studies so one can see whether formal analysis has been useful, where it has been useful, where it may need improvement. In this way we can begin to learn from different cultures and see ways in which we can pick up some interesting strands. However, we should not expect to emerge from these case studies with a grand synthesis. Rather, we should recognize that each problem, each situation, each culture, differs in certain specific ways that are extremely important in terms of where analysis comes into the picture. On that note I would like to suggest that we think about key research needs for IIASA which we can discuss further tomorrow.

Future Research Needs

The following question was put to the participants: What future research should IIASA undertake, bearing in mind IIASA's comparative advantages?:

- o problem-oriented research
- o international, comparative research
- o multidisciplinary backgrounds.

We have attempted below to categorize under broad headings the suggestions made by each of the four groups. We make no attempt to comment on the suggestions; they do provide food for thought regarding future directions that risk research might take, both at IIASA and at networking institutions. Finally, some of the suggestions may serve as a basis for future summer studies at IIASA.

1. Risk Analysis Issues

- o What is the role of social processes in the categorization of an issue?
- o What is the role of risk assessment?
- o How can methods for formal risk assessment be improved?
- o Develop simplified measures of risk for use in initial siting decisions (e.g., potential hazard zone for "credible" and "maximum" accident events in a fair/consistent manner).

Action-Oriented Risk Research

Seed project for action research:

Criteria: apolitical; multidisciplinary team of experts; far-reaching social/environmental/technical consequences.
Areas of application: Severn barrage tidal energy scheme (UK); LNG pipeline transportation (USSR); coal gasification (USA).

2. Institutional/Process Issues

- o Empirical validation criteria for decision analysis
- o Obstacles in moving to rational mode
- o Financial incentives in risk disputes (cross-national)
- o Legal/ethical aspects (cultural perspectives)
- o Adaptiveness of technology to changes in acceptance criteria
- o Comparative analysis of regulatory processes.

Areas of application: Drugs, DNA, medical technology, seat belts, acid rain, smoking, nuclear weapons, defense.

3. Issues Concerning International Comparative Risks

- o Different modes for incorporating risk analysis into the political process
- o Develop structure to facilitate comparison
- o What are the relative strengths and weaknesses of analysis?
- o Seek generic properties that cut across national/cultural differences
- o Styles of risk management appropriate for different politico-cultural systems
- o Comparison of methods already implemented (Berger Inquiry in Canada, Swedish remiss system, Dutch Science Shops)

4. Perceptions of Risk

- o Inadequacy of individual perception approach (imposed categories)
- o Focus on shared beliefs and goals and ideas about liability and equity
- o Sources of variance/extent of agreement between experts

5. Changing Attitudes of Society Towards Risk

- o Why are societies becoming increasingly more risk averse over time?
- o How do they decide what risks to be averse about?
- o What role does risk play in mental health?
- o Are we overconcerned with some risks and should we pay more attention to others (e.g., child abuse, deprived nations)?

6. Risks of War

- o What is the relationship between current defense, armaments and the risk of war?
  - — as involved with accumulating nuclear weapons
  - — as involved with sell-off of old arm stocks
  - — the functioning of threat systems
- o What is the impact of defense spending on national structure of other risks?

7. Issues in Conflict Resolution

- o Comparative studies of procedures for dispute resolution over technologies
- o Design of experimental studies on the group decision making process
- o Industrial labor relations practices in hazardous industries
- o Sociology of active impacted groups (cultural theory of risk)
- o Media and representation of images of risk
- o Sociology of experts and expertise

Appendix: List of Participants

Dr James Dooley
Institute of Environmental Studies
University of Toronto
Toronto 181
CANADA

Dr Mary Douglas
Russell Sage Foundation
633 3rd Avenue
New York, NY 10017
USA

Dr Elizabeth Drake
30F Inman Street
Cambridge, MA 02139
USA

Dr Tibor Englander
Institute of Psychology
Hungarian Academy of Sciences
Szondy u. 83-85
1068 Budapest
HUNGARY

Dr William Fairley
Analysis and Inference, Inc.
10 Post Office Square
Boston, MA 02109
USA

Professor Roger Kasperson
Center for Technology, Environment and Development
Clark University
Worcester, MA 01610
USA

Professor Paul Kleindorfer
Department of Decision Sciences
University of Pennsylvania
Philadelphia, PA 19104
USA

Professor Oleg Larichev
USSR Academy of Sciences
Committee for Systems Analysis
29 Ryleyev Street
Moscow 119034
USSR

Professor Dorothy Nelkin
Laboratoire d'Econometrie
Ecole Polytechnique
17 Rue Descartes
75230 Paris
FRANCE

Dr Yuri Oseredko
All-Union Designing Institute for Gas Industry
Kiev
USSR

Professor Howard Raiffa
Graduate School of Business Administration
Harvard University
Boston, MA 02163
USA

Dr Volker Ronge
Infratest Sozialforschung
Schluderstrasse 4
D-8000 Munich 19
FRG

Dr Werner Salz
Federal Ministry for Research and Technology
53 Bonn 2
FRG

Mr B. Schütz
Gesellschaft für Reaktorsicherheit (GRS) mbH
Forschungsgelände
8046 Garching
FRG

Dr Paul Schoemaker
Center for Decision Research
University of Chicago
Chicago, IL 60637
USA

Dr Brian Wynne
University of Lancaster
Bailrigg, Lancaster LA1 4YN
UK

IIASA PARTICIPANTS

Dr Roger Levien, Director
Mr Alec Lee, MMT Area Chairman
Dr Howard Kunreuther, Acting Task Leader, Risk Group
Mr Jesse Ausubel
Mr Hermann Atz
Dr John Lathrop
Ms Eryl Ley
Dr Joanne Linnerooth
Dr Chris Mandl
Dr Michael Stoto
Dr Michael Thompson

Index

MIX
Papier aus verantwortungsvollen Quellen
Paper from responsible sources
FSC® C105338

If you have any concerns about our products,
you can contact us on
ProductSafety@springernature.com

In case Publisher is established outside the EU,
the EU authorized representative is:
Springer Nature Customer Service Center GmbH
Europaplatz 3, 69115 Heidelberg, Germany

Printed by Libri Plureos GmbH
in Hamburg, Germany